# STRANGE STONES

ALSO BY PETER HESSLER

*Country Driving*

*Oracle Bones*

*River Town*

# STRANGE STONES

### DISPATCHES FROM EAST
### AND WEST

## PETER HESSLER

HARPER ⬤ PERENNIAL

NEW YORK • LONDON • TORONTO • SYDNEY • NEW DELHI • AUCKLAND

HARPER ⚫ PERENNIAL

FIRST EDITION

*Designed by William Ruoto*

Library of Congress Cataloging-in-Publication Data is available upon request.

ISBN 978-0-06-220623-7

13 14 15 16 17   OV/RRD   10 9 8 7 6 5 4 3 2 1

*For John McPhee*

# CONTENTS

WHEN I WAS GROWING UP, MY FATHER OCCASIONALLY BROUGHT his kids along to interviews. He was a medical sociologist at the University of Missouri, and his work took him to places that seemed exotic to me and my sisters: prisons, mental institutions, rural health clinics. Once, he met with the scion of an extended family that lived in the Mark Twain National Forest, deep in the Ozarks, where the clan was notorious for exerting heavy-handed control over surrounding hamlets. The old man's name was Elijah, and during the interview he sat next to an open window with a .22 rifle in his lap, in case a squirrel entered the conversation. He was eighty years old. When my father asked if there was a local drug problem, Elijah nodded seriously. "Yes, we do have a drug problem," he said. "There's no drugstore down here. If we need something, we have to drive all the way to Salem."

Elijah mentioned that a while back he had suffered throat pain so serious that he had trouble eating even watermelon, the local crop. Finally he went to a nearby town and saw a veterinarian, who performed a quick examination and diagnosed polyps. Elijah asked him to remove the growths.

"I'm sorry, but I'm not a medical doctor," the veterinarian said. "I can't do that."

"Well, mebbe you can't," Elijah said, "and mebbe you can."

And that was how it went—maybe nobody made a direct threat, and maybe Elijah didn't leave that office until the surgery was done.

My father was endlessly fascinated by the people he interviewed. Of course he enjoyed talking to the characters and the crazies, but he also had a deep interest in the quiet ones, the people who went about everyday lives with a sense of routine and decency. Missouri was like a foreign country to him and my mother. Both of them had grown up in Los Angeles, and they never expected to spend most of their lives in the Midwest. But they made it their home; my father spent years researching health care in rural communities, and my mother, a historian, wrote a dissertation about Jewish immigrants in Missouri.

My father can strike up a conversation with anybody. If a workman comes to the house, my father knows his life story by the time the project is finished. Once, a plumber repaired our bathroom, and he got along so well with my father that they still go deer hunting together in northern Missouri. During my childhood, if my father and I happened to be sitting somewhere with nothing to do—a bus station, a hotel lobby—he would pick out an individual and ask me what I noticed about him. Is there anything interesting about the way he's dressed? The way he carries himself? What do you think he does, why do you think he's here?

He had picked up this activity from his teacher in graduate school, a sociologist named Peter Kong-ming New. Peter New grew up in Shanghai but went to college in the United States; after the Communists took over in 1949, he stayed in America. He taught my father at the University of Pittsburgh, and then for a couple of years they worked together at Tufts, in Boston. Peter believed that I had been named after him, and although this wasn't exclusively true—my parents had other friends and relatives with the same name—it was true enough that they didn't disabuse him of the sense of honor. He was an unforgettable presence from my childhood. He stood well over six feet tall, with broad shoulders and a big belly. He had a huge bald head and a face as round as a mooncake. In addition to the way he liked to observe people, he developed a technique that he called "creative bumbling." If Peter was in a situation where he needed to get something done—placate a traffic cop, or get a table at a crowded restaurant—he suddenly became a *waiguoren*, a stranger in a strange land, and invariably people did everything they could to ease this confused and stuttering Chinaman along without further incident. Peter New had a booming voice, and he loved telling stories; like my father, he had the rare combination of being both loquacious and observant. He also had the exile's ability to make himself at home anywhere. That was my first impression of Chinese people—as a child I assumed they were all massive, charismatic figures. Whenever I heard the word "Shanghai," I pictured a city of giants.

Many years later, when I came to live in China, I realized how unusual Peter Kong-ming New really was. It wasn't just his size—it was the way he talked, the way he observed. Most Chinese tend to be wary of strangers, and there isn't a strong tradition of sociology and anthropology, of taking an interest in communities that are different from your own. In my experience, the Chinese aren't natural storytellers—they are often deeply modest, and they dislike being at the center of attention. After I began working as a journalist, I learned to be patient, because it often took months or even years to get a person to talk freely. I remembered my father's approach—if you want to truly understand somebody, you can't become bored or impatient, and the everyday matters as much as the exceptional. And there were many moments in China that called for some creative bumbling on the part of the *waiguoren*.

DESPITE THESE EARLY INFLUENCES, I DIDN'T PLAN TO BECOME a writer in China. Other than Peter New, I had no link to the region, and in college I never took a course in an Asian subject. My parents weren't the type to guide their children's careers. My father took me and my sisters to his interviews not because he hoped we would follow in his footsteps, but because he believed that life is more interesting if you can step outside of your own world every once in a while. And each sibling was encouraged to follow her own interests. Two of my sisters are married to policemen; one of them teaches, like my mother, and the other was a sociology major who is now a stay-at-home mom. My third sister is a sedimentary geologist.

For many years I hoped to become a fiction writer. To me, it seemed a higher calling than journalism; I loved the language of great novels and the storyteller's voice. In college, I majored in creative writing, focusing on short stories, but at the end of my junior year I took a seminar in nonfiction that was taught by John McPhee. He was the most exacting teacher I ever had—the margins of my papers bristled with comments written in a tight left-handed script. "You can't make a silk purse out of this," he wrote next to one bad sentence. When I loaded a phrase with adjectives and clauses, he responded, "This could be said with several pebbles removed from the mouth." In one profile, I used the subject's name four times in the span of two sentences, and McPhee wrote: "Listen to the character's name thudding like horseshoes. Vary it. Use pronouns here and there." He could be blunt: "This sort of thing is irritatingly repetitive." Another comment said simply, "This is lame cleverness."

But there were also marks of praise—"yes" or "ah" or "a fine moment." I realized that it's possible to write very well and very badly at the same time, and that the best writers aren't necessarily the most gifted but rather the ones who recognize their weaknesses and work hard to improve. By the end of the course I understood that nonfiction writing could be every bit as demanding as the work of a novelist. As time passed, I sensed that the routines of a fiction writer felt too internal for me, especially since I had a tendency toward shyness. I wanted work that forced me out; I needed contact with other lives, other worlds. This instinct inspired me to sign up for the Peace Corps, which sent me to

China. But the place itself was almost incidental—all I knew was that if I hoped to become a writer, I should go far from home.

THE STORIES IN THIS BOOK WERE WRITTEN BETWEEN THE YEARS 2000 and 2012. I wrote the first one when I was thirty years old, and over the next decade my life went through many changes. For a number of years I was single; then I was married, and finally (and all at once) a father of two after my wife and I had twin daughters. I lived in twelve different homes in three countries. A couple of these pieces were written in hotel rooms.

But this period taught me that writing could be an anchor. It felt the same wherever I went, and the basics of reporting—the curiosity, the patience, the willingness to connect with somebody different—had been introduced to me during childhood. After living overseas for a long time, and spending much of my life in another language, I tried to combine the perspectives of both the local and the outsider. Many of these stories are from China, where I lived for more than a decade, but there are also pieces from the United States, as well as Japan and Nepal. The essay that examines what it means to feel half-foreign, "Go West," is actually about returning to America.

I often wrote about people who were also on the move. I found myself attracted to the migrants and the transplants, the ones who were searching and the ones who were fleeing. I liked the folks who felt a little out of place. Some were chameleonlike, and others dreamed of returning home; a few engaged in various forms of creative bumbling. But they were all good to talk to, because they had learned to describe their surroundings with an outsider's eye.

These stories are not chronological, and only a few are about historical events: the closing of the Three Gorges Dam, the Beijing Olympics, the first peaceful transition of a national leader in Communist China. I've arranged them in this order for purely personal reasons, because I like the idea of David Spindler standing next to Rajeev Goyal, and I think that the people of Paradox might have something to say to the people of Wushan. Almost all of these pieces first appeared in *The New Yorker*, although they have been revised, often substantially, for this book. The magazine was another anchor during this period. I was fortunate to have the support of excellent editing and fact-checking, but mostly I appreciated the range of subjects and voices that the magazine was willing to publish. Out in the great wide world, foreign reporting can be depressingly narrow, especially in the post-9/11 climate. Sometimes it seems as if there are only two possible subjects for stories: people we should fear and people we should pity. But those aren't the individuals I met while living abroad.

It helped that *The New Yorker* allowed me to write about them in my own voice. One challenge for a foreign correspondent is to figure out how much of yourself to include: If a story is too self-centered, it becomes a tourist's diary. These days, the general trend is to reduce the writer's presence, often to the point of invisibility. This is the standard approach of newspapers, and it's described as a way of maintaining focus and impartiality. But it can make the subject feel even more distant and foreign. When I wrote about people, I wanted to describe the ways we interacted, the things we shared and the things that separated us. Chinese sometimes responded to me in certain ways because I was a *wai-guoren*, and it seemed important to let the reader know this. Mostly, though, I wanted to convey how things actually felt—the experience of living in a Beijing *hutong*, or driving on Chinese roads, or moving to a small town in rural Colorado. The joy of nonfiction is searching for balance between storytelling and reporting, finding a way to be both loquacious and observant.

But that's enough of that. China and Colorado are equally distant from me now; I've moved to a different country, with a new language to learn. Some days it feels overwhelming, and some days it feels like home.

*September 2012*
*Cairo, Egypt*

# WILD FLAVOR

"Do you want a big rat or a small rat?" the waitress says.

I'm getting used to making difficult decisions in Luogang. It's a small village in southern China's Guangdong Province, and I came here on a whim, having heard that Luogang has a famous rat restaurant. Upon arrival, however, I discovered that there are actually two celebrated restaurants—the Highest Ranking Wild Flavor Restaurant and the New Eight Sceneries Wild Flavor Food City. Both restaurants specialize in rat. They have the same bamboo and wood decor. They are next door to each other, and their owners are named Zhong and Zhong, respectively. Virtually everybody in Luogang village is named Zhong.

The restaurant Zhongs are not related, and the competition between them is keen. As a foreign journalist, I've been cajoled to such an extent that, in an effort to please both Zhongs, I agreed to eat two lunches, one at each restaurant. But before the taste test can begin, I have to answer the question that's been posed by the waitress at the Highest Ranking Wild Flavor Restaurant. Her name is Zhong. In Chinese it means "bell." She asks the question again: "Do you want a big rat or a small rat?"

"What's the difference?" I say.

"The big rats eat grass stems, and the small ones eat fruit."

This piece of information does not help much. I try a more direct tack. "Which tastes better?"

"Both of them taste good."

"Which do you recommend?"

"Either one."

I glance at the table next to mine. Two parents, a grandmother, and a little boy are having lunch. The boy is gnawing on a rat drumstick. I can't tell if the drumstick once belonged to a big rat or a small rat. The boy eats quickly. It's a warm afternoon. The sun is shining. I make my decision.

"Small rat," I say.

THE CHINESE CLAIM THAT FOLKS IN GUANGDONG WILL EAT anything. Besides rat, a customer at the Highest Ranking Wild Flavor Restaurant can order turtledove, fox, cat, python, and an assortment of strange-looking local animals whose names do not translate well into English. All of them are kept live in pens at the back of the restaurant, and they are killed only when ordered by a customer. Choosing among them is complicated, and it involves more than exoticism. You do not eat cat simply for the thrill of eating cat. You eat cat because cats have a lively *jingshen*, or spirit, and thus by eating the animal you will improve your spirits. You eat snake to become stronger. You eat deer penis to improve virility. And you eat rat to improve your—well, to be honest, I never knew that there was a reason for rat-eating until I got to Luogang, where every Zhong was quick to explain the benefits of the local specialty.

"It keeps you from going bald," said Zhong Shaocong, the daughter of the owner of the Highest Ranking Wild Flavor Restaurant.

"If you have white hair and eat rat regularly, it will turn black," said Zhong Qingjiang, who owns the New Eight Sceneries Wild Flavor Food City. "And if you're going bald and you eat rat every day your hair will stop falling out. A lot of the parents around here feed rat to a small child who doesn't have much hair, and the hair grows better."

Earlier this year, Luogang opened a "Restaurant Street" in the new Luogang Economic Development Zone, which is designed to draw visitors from the nearby city of Guangzhou. The government invested $1.2 million in the project, which enabled the two rat restaurants to move from their old, cramped quarters in a local park. On March 18, the Highest Ranking Wild Flavor Restaurant began serving customers in a twenty-thousand-square-foot facility that cost $42,000. Six days later, the New Eight Sceneries Wild Flavor Food City opened, with an investment of $54,000. A third restaurant—a massive, air-conditioned facility, which is expected to cost $72,000—will open soon. A fourth is in the planning stages.

"Their investment wasn't as much as mine," Deng Ximing, the owner of the third restaurant, told me. "You can see that my place is going to be much nicer. We have air-conditioning, which none of the others have."

It was early morning, and we were watching the workers lay the cement floor of the new restaurant. Deng was the only local restaurateur with a different family name, but he was married to a Zhong. He was in his midforties, and he had the fast-talking confidence of a successful entrepreneur. I also noticed that he had a good head of hair. He took great pride in Luogang Village's culinary tradition.

"It's more than a thousand years old," he said. "And it's always been rats from the mountains—we're not eating city rats. The mountain rats are clean, because up there they aren't eating anything dirty. Mostly they eat fruit—oranges, plums, jackfruit. People from the government hygiene department have been here to examine the rats. They took them to the laboratory and checked them out thoroughly to see if they had any diseases, and they found nothing. Not even the slightest problem."

Luogang's Restaurant Street has been a resounding success.

Newspapers and television stations have reported extensively on the benefits of the local specialty, and an increasing number of customers are making the half-hour trip from Guangzhou. Both the Highest Ranking Wild Flavor Restaurant and the New Eight Sceneries Wild Flavor Food City serve, on average, three thousand rats every day on the weekend. "Many people come from faraway places," Zhong Qingjiang told me. "They come from Guangzhou, Shenzhen, Hong Kong, Macao. One customer came all the way from America with her son. They were visiting relatives in Luogang, and the family brought them here to eat. She said you couldn't find this kind of food in America."

In America, you would also be hard-pressed to find twelve thousand fruit-fed rats anywhere on any weekend, but this isn't a problem in Luogang. On my first morning in the village, I watched dozens of farmers come down from the hills, looking to get a piece of the rat business. They came on mopeds, on bicycles, on foot; and all of them carried squirming burlap sacks full of rats that had been trapped on their farms.

"Last year I sold my oranges for fifteen cents a pound," a farmer named Zhong Senji told me. "But this year the price has dropped to less than ten cents." Like many other farmers, Zhong decided that the rat business was a lot better than the orange business. Today he had nine rats in his sack, which was weighed by a worker at the Highest Ranking Wild Flavor Restaurant. The bag shook and squeaked on the scale. It weighed in at just under three pounds, and Zhong received the equivalent in yuan of $1.45 per pound, for a total of $3.87. In Luogang, rats are more expensive than pork or chicken. A pound of rat costs nearly twice as much as a pound of beef.

At the Highest Ranking Wild Flavor Restaurant, I begin with a dish called Simmered Mountain Rat with Black Beans. The menu also includes Mountain Rat Soup, Steamed Mountain Rat, Simmered Mountain Rat, Roasted Mountain Rat,

Mountain Rat Curry, and Spicy and Salty Mountain Rat. But the waitress enthusiastically recommended the Simmered Mountain Rat with Black Beans, which arrives in a clay pot.

I eat the beans first. They taste fine. I poke at the rat meat. It's clearly well done, and it's attractively garnished with onions, leeks, and ginger. Nestled in a light sauce are skinny rat thighs, short strips of rat flank, and delicate toylike rat ribs. I start with a thigh, put a chunk of it into my mouth, and reach for a glass of beer. The beer helps.

The restaurant's owner, Zhong Dieqin, comes over and sits down. "What do you think?" she asks.

"I think it tastes good."

"You know it's good for your health."

"I've heard that."

"It's good for your hair and skin," she says. "It's also good for your kidneys."

Earlier this morning, I met a farmer who told me that my brown hair might turn black if I ate enough rat. Then he thought for a moment and said that he wasn't certain if eating rat has the same effect on foreigners that it does on the Chinese—it might do something entirely different to me. The possibility seemed to interest him a great deal.

At the table, Zhong Dieqin watches me intently. The audience also includes much of the restaurant staff. "Are you sure you like it?" asks the owner.

"Yes," I say, tentatively. In fact, it isn't bad. The meat is lean and white, without a hint of gaminess. There's no aftertaste. Gradually, my squeamishness fades, and I try to decide what the meat reminds me of, but nothing comes to mind. It simply tastes like rat.

After a while, Zhong Dieqin excuses herself, and the waitresses drift away. A young man comes over and identifies himself as the restaurant's assistant manager. He asks who I write for, and whether I came to Luogang specifically to report on the restaurants. None of my answers seems to satisfy him, and there's a wariness in his voice. I recognize it as a syndrome that's still pervasive in some parts of China: Fear of a Foreign Writer.

"Did you register with the government before you came here?" he asks.

"No."

"Why not?"

"Because it's too much trouble."

"You should have done that," he says. "Those are the rules."

"I don't think the government cares very much if I write about restaurants."

"They could help you," he says. "They would give you statistics and arrange interviews."

"I can find my own interviews. And if I registered with the government I would have to take all of the government officials out to lunch." A scene appears in my mind: a gaggle of Communist cadres, middle-aged men in cheap suits, all of them eating rat. I put my chopsticks down. The assistant manager keeps talking.

"A lot of foreigners come to our China to write about human rights," he says.

"That's true."

He looks at me hard. "Have you come here to write about human rights?"

"Have I asked you any questions about human rights?"

"No."

"Well, then, it would be hard for me to write a story about human rights."

He thinks about this for a while, but he still looks unsatisfied.

"I'm writing a story about Luogang's rat restaurants," I say. "It's nothing sensitive."

"You should have registered with the government," he says again. And I can see that if we keep talking he will repeat this phrase over and over, because this is one of those conversations that has been doomed by paranoia. It's a sad truth in China: even a perfectly good rat meal can be contaminated by politics.

I shrug and gather my things to leave, and the assistant manager requests that I not use his name in my article. I ask if I can use his family name.

"No," he says firmly.

"What's the risk?" I ask. "Everybody in Luogang has the same name anyway."

But his paranoia is particularly deep-rooted and he refuses. I thank him and promise that I won't mention his family name in my story. And I don't.

NEXT DOOR, AT THE NEW EIGHT SCENERIES WILD FLAVOR Food City, the Zhongs are far more media-savvy. They ask if I brought along a television crew.

"No," I say. "I don't have anything to do with television."

Zhong Qingjiang, the owner, is clearly disappointed. She tells me that a Hong Kong station came last month. She escorts me to a table, and the floor manager sits next to me. She asks, "How was the other restaurant?"

"It was fine," I say.

"What did you eat?"

"Simmered Mountain Rat with Black Beans."

"You'll like ours better," she says. "Our cook is better, the service is quicker, and the waitresses are more polite."

I decide to order the Spicy and Salty Mountain Rat. This time, when the waitress asks about size, I respond immediately. "Big rat," I say, pleased with my boldness.

"Come and choose it."

"What?"

"Pick out the rat you want."

In Chinese restaurants, fish and other seafood are traditionally shown live to customers for approval, as a way of guaranteeing freshness. It's not what I expected with rat, but now that the invitation has been made, it's too late to back out. I follow one of the kitchen workers to a shed behind the restaurant, where cages are stacked one on top of the other. Each cage contains more than thirty rats. The shed does not smell good. The worker points at a rat.

"How about this one?" he says.

"Um, sure."

He puts on a leather glove, opens the cage, and picks up the chosen rat. It's about the size of a softball. The rat is calm, perched on the hand of the worker, who keeps a grip on the tail.

"Is it OK?" asks the worker.

"Yes."

"Are you certain?"

The rat gazes at me with beady eyes. I have a strong desire to leave the shed.

"Yes," I say. "It's fine."

Before I go, the worker makes a sudden motion. He flips his wrist, keeps a grip on the tail, and swings his arm quickly. The rat makes a neat arc in the air. There is a soft thud when its head strikes the cement floor. There isn't much blood. The worker grins.

"Oh," I say.

"You can sit back down now," says the worker. "We'll bring it out to you soon."

Less than fifteen minutes later the dish is at my table. This time the chunks of rat are garnished with carrots and leeks. The chef comes out of the kitchen to join the rest of the audience, which consists of the owner, the floor manager, and a cousin of the owner. I take a bite.

"How is it?" the chef asks.

"Good."

"Is it too tough?"

"No," I say. "It's fine."

In truth, I'm trying hard not to taste anything. I lost my appetite in the shed, and now I eat quickly, washing every bite down with beer. I do my best to put on a good show, gnawing on the bones as enthusiastically as possible. When I finish, I sit back and manage a smile. The chef and the others nod with approval.

The owner's cousin says, "Next time you should try the Longfu Soup, because it contains tiger, dragon, and phoenix."

"What do you mean by 'tiger, dragon, and phoenix'?" I ask warily. I don't want to make another trip to the shed.

"It's not real tigers, dragons, and phoenixes," he says. "They're represented by other animals—cat for the tiger, snake for the dragon, and chicken for the phoenix. When you mix them together, there are all kinds of health benefits." He smiles and says, "They taste good, too."

# HUTONG KARMA

FOR THE PAST FIVE YEARS, I'VE LIVED ABOUT A MILE NORTH OF the Forbidden City, in an apartment building off a tiny alleyway in downtown Beijing. My alley has no official name, and it begins in the west, passes through three ninety-degree turns, and exits to the south. On a map, the shape is distinctive: it looks a little like a question mark, or perhaps half of a Buddhist swastika. The alley is also distinctive because it belongs to one of the few surviving sections of old Beijing. The capital, like all Chinese cities nowadays, has been changing fast—the biggest local map publisher updates its diagrams every three months, to keep pace with development. But the layout of my neighborhood has remained more or less the same for centuries. The first detailed map of Beijing was completed in 1750, under the reign of the great Qing dynasty emperor Qianlong, and on that document my alley follows the same route it does today. Xu Pingfang, a Beijing archaeologist, has told me that my street may very well date to the fourteenth century, when many sections of the city were originally laid out, under the Yuan dynasty. The Yuan also left the word *hutong*, a Mongolian term that has come to mean "alley" in Chinese. Locals call my alley Little Ju'er, because it connects with the larger street known as Ju'er Hutong.

I live in a modern three-story building, but it's surrounded by the single-story homes of brick, wood, and tile that are characteristic of *hutong*. These structures stand behind walls of gray brick, and often a visitor to old Beijing is impressed by the sense of division: wall after wall, gray brick upon gray brick. But actually a *hutong* neighborhood is most distinguished by connections and movement. Dozens of households might share a single entrance, and although the old residences have running water, few people have private bathrooms, so public toilets play a major role in local life. In a *hutong*, much is communal, including the alley itself. Even in winter, residents bundle up and sit in the road, chatting with their neighbors. Street vendors pass through regularly, because the *hutong* are too small for supermarkets.

There are few cars. Some alleys, like the one I live on, are too narrow for automobile traffic, and the sounds of daily life are completely different from what one would expect in the heart of a city of seven million people. Usually I'm awake by dawn, and from my desk I hear residents chatting as they make their way to the public toilet next to my building, chamber pots in hand. By midmorning the vendors are out. They pedal through the alley on three-wheeled carts, each announcing his product with a trademark cry. The beer woman is the loudest, singing out again and again: "*Maaaaiiii piiiiijiuuuuuu!*" At eight in the morning, it can be distracting—buuuuyyyy beeeecccceeer!—but over the years I've learned to appreciate the music in the calls. The rice man's refrain is higher-pitched; the vinegar dealer occupies the lower scales. The knife-sharpener provides percussion—a steady click-clack of metal plates. The sounds are soothing, a reminder that even if I never left my doorway again life would be sustainable, albeit imbalanced. I would have cooking oil, soy sauce, and certain vegetables and fruit in season. In winter I could buy strings of garlic. A vendor of toilet paper would pedal through every day. There would be no shortage of coal. Occasionally I could eat candied crabapple.

I could even earn some money from the freelance recyclers.

On an average day, a recycler passes through every half hour, riding a flatbed tricycle. They purchase cardboard, paper, Styrofoam, and broken appliances. They buy old books by the kilogram and dead televisions by the square inch. Appliances can be repaired or stripped for parts, and the paper and plastic are sold to recycling centers for the barest of profits: the margins of trash. Not long ago, I piled some useless possessions in the entranceway of my apartment and invited each passing recycler inside to see what everything was worth. A stack of old magazines sold for sixty-two cents; a burned-out computer cord went for a nickel. Two broken lamps were seven cents, total. A worn-out pair of shoes: twelve cents. Two broken Palm Pilots: thirty-seven cents. I gave one man a marked-up manuscript of the book I'd been writing, and he pulled out a scale, weighed the pages, and paid me fifteen cents.

One day in late April, I was sitting at my desk and heard somebody call out, "Looonnnng haaaaiiiir! Looonnnng haaaaiiiir! Looonnnng haaaaiiiir!" That was a new strain of music, so I went out into the alley, where a man had parked his cart. He had come from Henan Province, where he worked for a factory that produces wigs and hair extensions. When I asked about business, he reached inside a burlap sack and pulled out a long black ponytail. He said he'd just bought it from another *hutong* resident for ten dollars. He had come to Beijing because it was getting warm—haircut season—and he hoped to acquire one hundred pounds of good hair before returning to Henan. Most of it, he said, would eventually be exported to the United States or Japan.

While we were talking, a woman hurried out of a neighboring house, carrying something in a purple silk handkerchief. Carefully, she unwrapped it: two thick strands.

"They're from my daughter," she said, explaining that she'd saved them from the last haircut.

Each ponytail was about eight inches long. The man held

one up so that it dangled like a fish on the line. He squinted appraisingly and said, "Those are too short."

"What do you mean?"

"They're no use to me," he said. "They need to be longer than that."

The woman tried to bargain, but she didn't have much leverage; finally she returned home, hair in hand. The call echoed as the man left the *hutong*: "Looonnnng haaaaiiiir! Looonnnng haaaaiiiir!"

NOT LONG AFTER I MOVED INTO LITTLE JU'ER, BEIJING stepped up its campaign to host the 2008 Games, and traces of Olympic glory began to touch the *hutong*. In order to boost the athleticism and health of average Beijing residents, the government constructed hundreds of outdoor exercise stations. The painted steel equipment is well-intentioned but odd, as if the designer had caught a fleeting glimpse of a gym and then worked from memory. At the exercise stations, people can spin giant wheels with their hands, push big levers that offer no resistance, and swing on pendulums like children at a park. In the greater Beijing region, the stations are everywhere, even in tiny farming villages by the Great Wall. Out there, the equipment gives the peasants a new lifestyle option: after working a twelve-hour day on the walnut harvest, they can get in shape by spinning a big yellow wheel over and over.

But nobody appreciates the exercise stations more than *hutong* residents. The machines are scattered throughout old parts of the city, tucked into narrow alleyways. At dawn and dusk, they are especially busy—older people meet in groups to chat and take a few rounds on the pendulum. On warm evenings, men sit idly on the machines and smoke cigarettes. The workout stations are perfect for the ultimate *hutong* sport: hanging around in the street with the neighbors.

At the end of 2000, as part of the citywide pre-Olympic

campaign to improve sanitation facilities, the government re-built the public toilet at the head of Ju'er Hutong. The change was so dramatic that it was as if a shaft of light had descended directly from Mount Olympus to the alleyway, leaving a magnificent structure in its wake. The building had running water, infrared-automated flush toilets, and signs in Chinese, English, and Braille. Gray rooftop tiles recalled traditional *hutong* architecture. A list of detailed rules was printed onto stainless steel: "Number 3: Each user is entitled to one free piece of common toilet paper (length 80 centimeters, width 10 centimeters)." A small room housed a married couple who served as full-time attendants. Realizing that no self-respecting Beijing resident would work in a public toilet, the government had imported dozens of couples from the interior, mostly from the poor province of Anhui. The logic was sound: the husband cleaned the men's room, the wife cleaned the women's.

The couple in Ju'er Hutong brought their young son, who took his first steps in front of the public toilet. Such scenes occurred across the capital, and perhaps someday the kids would become the Beijing version of Midnight's Children: a generation of toddlers raised in public toilets who, ten years after the Olympics, will come of age and bring hygienic glory to the Motherland. Meanwhile, Ju'er residents took full advantage of the well-kept public space that fronted the new toilet. Old Yang, the local bicycle repairman, stored his tools and extra bikes there, and in the fall cabbage vendors slept on the strip of grass that bordered the bathroom. Wang Zhaoxin, who ran the cigarette shop next door, arranged some ripped-up couches around the toilet entrance. Someone else contributed a chessboard. Folding chairs appeared, along with a wooden cabinet stocked with beer glasses.

After a while, there was so much furniture, and so many people there every night, that Wang Zhaoxin declared the formation of the "W. C. Julebu": the W. C. Club. Membership was open to all, although there were disputes about who should be chairman or a member of the Politburo. As a foreigner, I joined at the level of a Young Pioneer. On weekend nights, the club hosted barbecues in front of the toilet. Wang Zhaoxin supplied cigarettes, beer, and grain alcohol, and Mr. Cao, a driver for the Xinhua news service, discussed what was happening in the papers. The coal-fired grill was attended to by a handicapped man named Chu. Because of his disability, Chu was licensed to drive a small motorized cart, which made it easy for him to transport skewers of mutton through the *hutong*. In the summer of 2002, when the Chinese men's soccer team made history by playing in its first World Cup, the W. C. Club acquired a television, plugged it into the bathroom, and mercilessly mocked the national team as it failed to score a single goal throughout the tournament.

WANG ZHAOXIN MODESTLY REFUSED THE TITLE OF CHAIRMAN, although he was the obvious choice, because nobody else had seen so many changes in the neighborhood. Wang's parents had moved to Ju'er Hutong in 1951, two years after the Communist Revolution. Back then, Beijing's early-fifteenth-century layout was still intact, and it was unique among major world capitals: an ancient city virtually untouched by modernity or war.

Beijing had once been home to more than a thousand temples and monasteries, but nearly all of them were disbanded and converted to other uses by the Communists. In Ju'er, the monks were kicked out of a lamasery called Yuan Tong Temple, and dozens of families moved in, including Wang Zhaoxin's parents. Meanwhile, other members of the proletariat were encouraged to occupy the homes of the wealthy. Previously, such private *hutong* residences had been arranged around spacious open-air courtyards, but during the 1950s and 1960s most of these became crowded with shanties and makeshift structures. The former compound of a single clan might become home to two dozen families, and the city's population swelled with new arrivals. Over the next twenty years, the Communists tore down most of Beijing's monumental gates, as well as its impressive city wall, which in some places was forty feet high. In 1966, when Wang Zhaoxin was a six-year-old elementary-school student, he participated in a volunteer children's work brigade that helped demolish a section of the Ming dynasty city wall not far from Ju'er. In 1969, during the Cultural Revolution, the nearby Anding Gate was torn down to make room for a subway station. By the time Mao died, in 1976, roughly a fifth of old Beijing had been destroyed.

In 1987, Wang Zhaoxin's younger brother accepted his first job, at a Beijing noodle factory. Within months of starting work, the eighteen-year-old lost his right arm in a flour-mixing machine. Not long before that, Wang Zhaoxin had decided to go into retail, hoping to succeed in the new market economy, and now he chose a product line in deference to his brother's disability. Fruit and vegetables are too heavy, he reasoned, and a clothes merchant needs two arms to measure and fold goods. Cigarettes are light, so that's what the Wang brothers sold.

During the 1990s and early 2000s, as the Wangs hawked cigarettes in Ju'er Hutong, developers sold most of old Beijing. Few sections of the city were protected, in part because local government bureaus profited from development. Whenever a *hutong* was doomed, its buildings were marked with a huge painted character surrounded by a circle, like the "A" of the anarchist's graffiti:

*Chai*: "Pull down, dismantle." As developers ran rampant over the city, that character became a talisman—Beijing artists riffed on the shape, and residents cracked *chai* jokes. At the W. C. Club, Wang Zhaoxin used to say, "We live in *Chai nar*." It sounded like the English word "China," but it meant "Demolish where?"

Like many Beijing people I knew, Wang Zhaoxin was practical, good-humored, and unsentimental. His generosity was well known—locals had nicknamed him Wang Laoshan, Good Old Wang. He always contributed more than his share to a W. C. Club barbecue, and he was always the last to leave. He used to say that it was only a matter of time before the government *chai*'d more buildings in our neighborhood, but he never dwelled on the future. More than four decades in *Chai nar* had taught him that nothing lasts forever.

The W. C. Club was near the head of the *hutong*, which ends at Jiaodaokou South Street. That boulevard is busy with streetcars and buses; the nearest intersection is home to a massive new apartment complex, two supermarkets, and a McDonald's. Jiaodaokou represents a border: by stepping onto the street, you enter the modern city.

Every day, most working residents of the *hutong* cross that divide. They pass the bicycle-repair stand of Old Yang, who keeps his pumps and toolbox next to the Olympic toilet. In a *hutong*, there's no better network than one that combines bikes and bathrooms, and Old Yang knows everybody. Occasionally he gives me messages from other people in the neighborhood; once he passed along the business card of a foreigner who had been trying to track me down. Another time he told me that the local matchmaker had somebody in mind for me.

"College-educated, 1.63 meters tall," he said curtly. Those were the only specs he knew. For Chinese women, 1.6 is a magic number—you often see that figure in job listings and dating ads. It's about five feet three. I told Old Yang that I appreciated the tip but that I didn't want to meet anybody right now.

"Why not? You're not married."

"Well, I'm not in a rush. In my country people get married later."

When I started to walk off, he told me that he'd already given my phone number to the matchmaker.

"Why did you do that?" I said. "You have to tell her that I'm not interested."

Old Yang is in his sixties, a tall, stern-faced man with a shaved head. When I tried to decline the invitation, his expression became even more serious than usual. He told me it was too late: everything had already been arranged, and he told me he'd look bad if I didn't go. That week, the matchmaker called me four times. She introduced herself as Peng Laoshi—Teacher Peng—and she had scheduled the date for Saturday afternoon. We met beyond the *hutong*'s boundary, at the entrance to the Jiaodaokou McDonald's. My date was supposed to arrive in a few minutes, but there was something that Teacher Peng wanted to clarify first.

"This is an underground meeting," she said, after we had found seats in the upstairs section of the restaurant.

"Why?"

"It's not official. We're not allowed to work with foreigners."

"Why not?"

"The government doesn't want us to," she said. "They're afraid the foreigners will trick Chinese women."

There was a pause, from which point the conversation could have proceeded in any number of interesting directions. But Teacher Peng seemed accustomed to filling awkward silences. "Of course, I'm not worried about you," she said, beaming and speaking fast. "Old Yang says you're a good person."

Teacher Peng was in her midforties, and the skin around her eyes was crinkled from smiling so much—a rare characteristic in China. She wasn't an actual teacher; that's simply the title people use for matchmakers. In China, professional matchmakers still play a role in rural areas and small cities, but they've become less important in places like Beijing. Nevertheless, I occasionally see a sign advertising their services, especially in old neighborhoods. Teacher Peng kept a government-registered office in Ju'er.

At McDonald's, I asked Teacher Peng how much she charged, and she said the fee for meeting someone was usually two hundred yuan.

"But it's more to meet a foreigner," she said. "Five hundred, one thousand, even two thousand."

I asked, as delicately as possible, how much today's client would have to pay for me if things worked out.

"One thousand." It was a little more than $120. Even if other foreigners were worth twice as much, there was some consolation in being double the minimum.

"Does she have to pay anything just for meeting today?" I asked.

"No. It's only if you stay together."

"For marriage?"

"No. For more dates."

"How many?"

"That depends."

She wouldn't give me a number, and I kept asking questions, trying to figure out how the system worked. At last she leaned forward and said: "Do you hope to get married quickly, or do you just want to spend time with a woman?"

It was a hell of a first-date question for a single male in his early thirties. What could I say? I didn't want the bike repairman to lose face. "I really don't know," I stammered. "But I want to make sure that she's not paying anything to meet me today."

Teacher Peng smiled again. "You don't have to worry about that," she said.

WHEN I FIRST MOVED TO THE NEIGHBORHOOD, I REGARDED McDonald's as an eyesore and a threat: a sign of the economic boom that had already destroyed most of old Beijing. Over time, though, *hutong* life gave me a new perspective on the franchise. For one thing, it's not necessary to eat fast food in order to benefit from everything that McDonald's has to offer. At the Jiaodaokou restaurant, it's common to see people sitting at tables without ordering anything. Invariably, many are reading; in the afternoon, schoolchildren do their homework. I've seen the managers of neighboring businesses sitting quietly, balancing their account books. And always, always, always somebody is sleeping. McDonald's is the opposite of *hutong* life, in ways both good and bad: cool in summer, warm in winter, with private bathrooms.

It's also anonymous. Unlike Chinese restaurants, where waitresses hover, the staff at a fast-food joint leaves people alone. On a number of occasions, dissidents have asked me to meet them at a McDonald's or a KFC, because it's safe. When Teacher Peng told me that our meeting was "underground," I realized why she had chosen the restaurant.

Others apparently had the same idea. One couple sat near the window, leaning close and whispering. At another table, two well-dressed girls seemed to be waiting for their dates. Over Teacher Peng's left shoulder, I kept an eye on a couple who appeared to be having some sort of crisis. The woman was about twenty-five; the man seemed older, in his forties. Their faces shone with the unnatural redness that comes to many Chinese who have been drinking. They sat in silence, glaring at each other. Nearby, the McDonald's Playland was deserted. Teacher Peng's pager went off.

"That's her," she said, and asked to borrow my cell phone.

"I'm at McDonald's," she said into the phone. "The Italian is already here. Hurry up."

After Teacher Peng hung up, I tried to say something, but she spoke too fast. "She teaches music at a middle school," she said. "She's a very good person—I wouldn't introduce you otherwise. Good. Listen. She's twenty-four years old. She's pretty and she's

1.64 meters tall. She's educated. She's thin, though. I hope that's not a problem—she's not as voluptuous as the women in your Italy."

There was so much to process—for one thing, my date seemed to be growing taller—but before I could speak Teacher Peng rattled on: "Good. Listen. You have a good job and you speak Chinese. Also, you were a teacher before, so you have something in common."

Finally she stopped to breathe. I said, "I'm not Italian."

"What?"

"I'm American. I'm not Italian."

"Why did Old Yang tell me you're Italian?"

"I don't know," I said. "My grandmother is Italian. But I don't think Old Yang knows that."

Now Teacher Peng looked completely confused.

"America is an immigrant country," I began, and then I decided to leave it at that.

She recovered her poise. "That's fine," she said with a smile. "America is a good country. It's fine that you come from America."

THE WOMAN ARRIVED WEARING HEADPHONES. JAPANESE SCRIPT decorated her stylish jacket, and she wore tight jeans. Her hair had been dyed a dark brown. Teacher Peng introduced us, crinkled her eyes one last time, and tactfully took her leave. Very slowly, one by one, the woman removed her headphones. She looked quite young. The CD player sat on the table between us.

I said, "What are you listening to?"

"Wang Fei"—a popular singer and actress.

"Is it good?"

"It's OK."

I asked her if she wanted anything from the restaurant, and she shook her head. I respected that—why spoil a date at McDonald's by eating the food? She told me that she lived with her parents in a *hutong* near the Bell Tower; her school was nearby. While she was talking, I glanced at the drunk couple behind her. Now they were ignoring each other and the woman flipped through a newspaper, angrily.

The music teacher said, "Do you live near here?"

"I live in Ju'er Hutong."

"I didn't know there were foreigners there," she said. "How much is your rent?"

This being China, I told her.

"That's a lot," she said. "Why do you pay so much?"

"I don't know. I guess they can always charge foreigners more."

"You were a teacher, right?"

I told her that I used to teach English in a small city in Sichuan Province.

"That must have been boring," she said. "Where do you work now?"

I said that I was a writer who worked at home.

"That sounds even more boring," she said. "I'd go nuts if I had to work at home."

The drunk couple began arguing loudly. Suddenly the woman stood up, brandished the newspaper, and smacked the man on the head. Then she spun on her heel and stormed out past Playland. Without a word, the man folded his arms, lay his head down on the table, and went to sleep.

The music teacher looked up at me and said, "Do you often go back to your Italy?"

THE FOLLOWING WEEK, THE MATCHMAKER TELEPHONED TO SEE IF there was any chance of a second meeting, but she wasn't insistent. She impressed me as a sharp woman—sharp enough to recognize that my cluelessness might be exploited in better ways than dates at McDonald's. The next time I ran into her in the *hutong*, she asked if I wanted to become an investor in a karaoke parlor. After that, I avoided walking past her office.

When I asked Old Yang about the confusion, he shrugged and said that I once mentioned that my grandmother is of Italian descent. I had no memory of the conversation, but I picked up a valuable *hutong* lesson: never underestimate how much the bike repairman knows.

GOOD OLD WANG WAS RIGHT ABOUT *CHAI NAR*. FOR YEARS, he had predicted demolition, and in September of 2005, when the government finally condemned his apartment building, he moved out without protest. He had already sold the cigarette shop, because the margins had fallen too low. And now there was no doubt who had been the true chairman, because the W. C. Club died as soon as he left the *hutong*.

By then, three-quarters of old Beijing had been torn down. The remaining quarter consisted mostly of public parks and the Forbidden City. Over the years, there had been a number of protests and lawsuits about the destruction, but such disputes tended to be localized: people complained that government corruption reduced their compensation, and they didn't like being resettled to suburbs that were too distant. But it was unusual to hear a Beijinger express concern for what was happening to the city as a whole. Few spoke in terms of architectural preservation, perhaps because the Chinese concept of the past isn't closely linked to buildings, as it is in the West. The Chinese rarely built of stone, instead replacing perishable materials periodically over the centuries.

The essence of the *hutong* had more to do with spirit than structure: it wasn't the brick and tiles and wood that mattered; it was the way that people interacted with their environment. And this environment had always been changing, which created residents like Good Old Wang, who was pragmatic, resourceful, and flexible. There was no reason for such people to feel threatened by the initial incursions of modernity—if anything, such elements tended to draw out the *hutong* spirit, because residents immediately found creative ways to incorporate a McDonald's or an Olympic toilet into their routines. But such flexibility could also make people passive when the incursions turned into wholesale destruction. That was the irony of old Beijing: the most appealing aspects of the *hutong* character helped pave the way for its destruction.

In 2005, the Beijing government finally instituted a new plan to protect the scattered old neighborhoods that remained in the north and west of downtown, including Ju'er. These *hutong* wouldn't be put on the market for developers to build whatever they wished, as they would have been in the past. The stated priority was to "preserve the style of the old city," and the government established a ten-member advisory board to consult on major projects. The board's members included architects, archaeologists, and city planners, some of whom had publicly criticized the destruction. One board member told me that it was essentially too late, but that the new plan should at least preserve the basic layout of the few surviving *hutong*. Within that layout, however, gentrification was inevitable—the *hutong* had become so rare that they now had cachet in the new economy.

The change happened fast in my neighborhood. In 2004, bars, cafés, and boutiques started moving into Nan Luogu Xiang, a quiet street that intersects Ju'er. Locals were happy to give up their homes for good prices, and the businesses maintained the traditional architectural style, but they introduced a new sophistication to the Old City. Nowadays, if I'm restricted to my neighborhood, I have access to Wi-Fi, folk handicrafts, and every type of mixed drink imaginable. There is a nail salon in the *hutong*. Somebody opened a tattoo parlor. The street vendors and recyclers are still active, but they have been joined by troops of pedicab men who give "*hutong* tours." Most of the tourists are Chinese.

One recent weekend, Good Old Wang returned for a visit, and we walked through Ju'er. He showed me the place where he was born. "There's where we lived," he said, pointing at the modern compound of the Jin Ju Yuan Hotel. "That used to be the temple. When my parents moved in, there was still one lama left."

We continued east, past an old red door that was suspended in the *hutong*'s wall, three feet above the street. "There used to be a staircase there," he explained. "When I was a child, that was an embassy."

In the nineteenth century, the compound had belonged to a Manchu prince; in the 1940s, Chiang Kai-shek used it as his Beijing office; after the revolution, it was taken over by Dong Biwu, a founder of the Chinese Communist Party. In the 1960s, it served as the Yugoslavian embassy. Now that all of them were gone—Manchus, Nationalists, Revolutionaries, Yugoslavians— the compound was called, appropriately, the Friendship Guest House.

That was *hutong* karma—sites passed through countless incarnations, and always the mighty were laid low. A couple of blocks away, the family home of Wan Rong, empress to the last monarch of the Qing dynasty, had been converted into a diabetes clinic. In Ju'er, the beautiful Western-style mansion of Rong Lu, a powerful Qing military official, had served one incarnation as the Afghanistan embassy before becoming what it is today: the Children's Fun Publishing Co., Ltd. A huge portrait of Mickey Mouse hangs above the door.

Good Old Wang passed the Olympic toilet ("it's a lot less cluttered than when I was here"), and then we came to the nondescript three-story building where he had lived since 1969. It wasn't a historic structure, which was why it had been approved for demolition. The electricity and the heat had been cut off; we walked upstairs into an abandoned hallway. "This was my room when I was first married," he said, stopping at a door. "Nineteen eighty-seven."

His brother had lost his arm that year. We continued down

the hall, to the apartment where Wang had lived most recently, with his wife, his daughter, his father, and his brother. The girl's drawings still decorated the walls: a sketch of a horse, the English phrase "Merry Christmas." "This is where the TV was," he said. "That's where my father slept. My brother slept there."

The family had since dispersed. The father and brother now live in a *hutong* to the north; Good Old Wang, his wife, and their daughter are using the home of a relative who is out of town. As compensation for the condemned apartment, Good Old Wang was given a small section of a decrepit building near the Drum Tower. He hoped to fix it up in the spring.

Outside, I asked him if it had been hard to leave the *hutong* after nearly half a century. He thought for a moment. "You know, lots of things happened while I lived here," he said. "And maybe there were more sad events than happy events."

We headed west out of the *hutong*. On the way, we passed an ad for the Beijing Great Millennium Trading Co., Ltd. Later that day, returning home, I saw a line of pedicabs: Chinese tourists, bundled against the cold, cameras in hand as they cruised the ancient street.

# WALKING THE WALL

WHEN THE WEATHER IS GOOD, OR WHEN I'M TIRED OF HAVING seven million neighbors, I drive north from downtown Beijing. It takes an hour and a half to reach Sancha, a quiet village where I rent a farmhouse. The road switchbacks up a steep hillside and dead-ends at the village, but a footpath continues into the mountains. The trail forks twice, climbs for a steep mile through a forest of walnut and oak, and finally terminates at the Great Wall of China.

Once, I packed a tent and sleeping bag, hiked up from the village, and walked eastward along the wall for two days without seeing another person. Tourists rarely visit this area, where the wall is perched high on a ridgeline, magnificent in its isolation. The structure is made of brick and mortar; there are crenellations, and archer slits, and guard towers that rise more than twenty feet high. The tallest one is known locally as the Great Eastern Tower, and it looms above a stretch of wall that contains an inscribed marble tablet. Originally, there were many such tablets, but this is one of fewer than ten that are known to remain on the wall in the Beijing region. The inscription notes that in 1615 A.D. a crew of 2,400 soldiers built a section of the wall that measured exactly fifty-eight *zhang* and five *cun*. There are one hundred *cun* in a *zhang*, and each *cun* is about an inch and a half; the total length of this wall section is around six hundred and fifty feet. The bureaucratic precision of the inscription, in this forgotten place, seems as lonely as words can be.

In November, I hiked to the Great Eastern Tower with two friends who were visiting from New York. After reaching the tower, we began the long descent to the south. This stretch can be treacherous; many brick ramparts have collapsed. I was picking my way downhill when something in the rubble caught my eye. It was white—too white to be brick, too big to be mortar. I dug it out and saw four neat rows of carved characters.

It was a fragment of another marble tablet. I could make out a few words: something was six *chi* high, something else was two *zhang*. But the writing was in classical Chinese, which I've never studied, and the surface was badly scarred.

"How long do you think it's been buried here?" one of my friends asked.

"I have no idea," I said. "But I think we better hide it."

We covered the fragment with loose bricks. I memorized the surrounding details, so I could find it again. A month later I returned with David Spindler.

DAVID SPINDLER STANDS SIX FEET SEVEN, AND HE IS RESERVED in a way that is characteristic of many men who are very tall and thin. He once remarked to me that height is the only physical trait that Americans comment on openly, making offhand remarks and jokes that can be rude. After that, I started noticing that whenever I saw Spindler at social gatherings, he had usually found a way to sit down. Beijing is full of foreigners who cultivate an air of eccentricity, but everything about Spindler seems designed to avoid attention. He rarely talks in detail about his research, and he doesn't promote himself as an expert. He chooses his words very carefully. He is thirty-nine years old, with short sandy hair, a long face, and gentle eyes. For people who know him casually in the city, the way I did for years, it's a surprise to witness the transformation that occurs when he visits the mountains.

On a cold December morning, Spindler and I drove to Sancha and set off on foot to find the marble tablet. He wore a red-checked wool hunting shirt, a floppy white Tilley safari hat, and high-end La Sportiva mountaineering boots. For a face mask, he'd cut a leg off a pair of sweatpants, scissored a round hole, and pulled it over his head. His polyurethane-coated L.L.Bean hunting trousers had been reinforced by a local street tailor—cheap denim patches covered the expensive pants, like a friendship quilt linking Freeport and Beijing. His hands were protected by huge elk-leather gloves designed for utility line workers by J. Edwards of Chicago. Spindler looked like a scarecrow of specialty gear—some limbs equipped for hard labor, others for intense recreation. Over the years, he had determined that this was precisely the right ensemble for the Great Wall, where thorns and branches are common.

We followed the wall east. Every hundred yards or so, it connected to a tower. These structures were crumbling but still impressive, with high vaulted ceilings and arched windows. Periodically, Spindler pointed out details: a place where a door used to be barred, a brick frame that had once contained an inscribed tablet.

"The towers and the wall were totally different projects," he said. "First, you had the brick towers, and the wall was just local stone. And then they came and improved the wall. That's why these towers look a little funny."

He pointed out a place where the wall's crenellation ran right into the open window of a tower—the kind of thing that happens when you use two different contractors. Near the Great Eastern Tower, one section of wall had collapsed entirely. Spindler believed that the construction project of 1615 had ended right there, at the edge of a short precipice. He had measured it once, using the details found on the marble tablet near the tower. "These guys really hosed the next construction crew," he said, gazing down at the precipice. "What could they do? It's really hard to build from that point."

I had hiked this section perhaps fifty times, but I had never noticed these details of construction. In my mind, it was simply the Great Wall—complete and virtually timeless. For Spindler, though, it was a work of many pieces and seasons. Construction generally took place in the spring, when the weather was good but Mongol raiders weren't active. "Energy in the Mongol world was fat on the horses," he said. "They didn't have that after the winter, so the spring was not a good season for raiding. Summer was too hot. They didn't like the heat; they didn't like the insects. The Mongol bowstrings were made of hide, and with the humidity they supposedly went slack—this is described in Ming texts. Most raids took place in the fall."

We came to the place where I had hidden the broken tablet. Spindler crouched in the cold, running a finger along the carved characters. He recognized it immediately as a piece of a tablet that dated to 1614. The county antiquities bureau had recorded its inscription in 1988, but they had failed to track down its original location on the wall, and since then it had disappeared. Some looter had probably broken it.

"It's saying the height of the wall, including the crenellations," Spindler explained. "And then it starts in with all the names of the officials. God, it's good that somebody got this down before it was destroyed."

He pulled a tape measure out of his backpack. After examining the space between the carved lines, he quickly calculated the original dimensions. Slowly, he walked back along the wall, looking for a place where it could have been mounted. He found an empty brick-bordered ledge and measured it: perfect fit. For this small section of the wall, he now knew the basic story of two construction campaigns in the 1610s. Before leaving, we returned the fragment to the spot where I had found it, covering the marble with broken bricks.

While we were there, a local farmer hiked up from the south. He was trapping game: a dozen wire snares were looped over his shoulder. If the presence of a six-foot-seven foreigner in a Tilley hat and utility-line elk-leather gauntlets surprised the farmer, he didn't show it. He asked if we had extra water, and Spindler gave him a bottle. During the following year, Spindler and I hiked through a number of villages together, and each time the locals hardly seemed to distinguish between the two of us. Andrew Field, a friend of Spindler's who teaches Chinese history at the University of New South Wales, once told me that an unusually tall person might actually feel more comfortable on the Great Wall than in America. "In China, sure, he's a monster," Field said. "But aren't we all?"

THE FIRST KNOWN HISTORICAL REFERENCE TO WHAT WE THINK of as the Great Wall dates to 656 B.C., during the Warring States period, when the kingdom of Chu constructed defensive barriers of packed earth. More than four centuries later, the state of Qin conquered all of its rivals, consolidating power across the north of what is now China. In 221 B.C., Qin Shihuang became the first man to declare himself emperor. After seizing power, he commanded the construction of roughly three thousand miles of *changcheng*.

The term translates literally as either "long wall" or "long walls"—Chinese doesn't differentiate between singular and plural—and Qin's barriers, like those of the Chu, consisted of hard-packed earth. Over the centuries, many dynasties faced the same basic problem: the wide-open frontier of the northern plains made the empire vulnerable to the nomadic Mongol and Turkic tribes that inhabited these lands. At times, the nomadic threat became more intense, and different dynasties responded with different strategies. The Tang, who ruled from 618 to 907 A.D., built virtually no walls, because the imperial family was part Turkic and skilled in Central Asian warfare and diplomacy. Even when dynasties constructed walls, they didn't necessarily call them *changcheng*—over the centuries, more than ten terms were used to describe the fortifications.

The Ming usually called theirs *bianqiang*—"border wall(s)"—and they became the greatest wall builders in Chinese history. They came to power in 1368, after the collapse of the Yuan, a short-lived

Mongol dynasty that had been founded by Kublai Khan. Even after the Mongols lost power in China, they continued to pose a threat in the north, and in the 1500s the Ming began to construct large fortifications of quarried stone and brick in the Beijing region. These are the iconic structures (some of them rebuilt and restored) that seem to continue endlessly in tourist photographs. The Ming was the only dynasty to build extensively with such durable materials, and many sections of wall ran for miles. But the *bianqiang* was a network rather than a single structure, and some regions had as many as four distinct lines of fortifications.

In 1644, domestic rebels stormed the capital, and the Ming emperor committed suicide. In desperation, a military commander in the northeast opened a major *bianqiang* gate to the Manchus, a northern tribe, in the hope that they would restore the ruling family. Instead, the Manchus founded their own dynasty, the Qing, which lasted until 1912. The Qing had little use for the walls—after all, these were people who had originally come from the far side of the barriers—and they abandoned the defensive system to the elements.

But as Western explorers and missionaries began to penetrate China in the eighteenth century, they toured the Ming ruins and confused them with stories they had been told about Qin Shihuang's three-thousand-mile wall. Foreigners assumed that the Beijing region's trellised brick fortifications were part of an unbroken line that had stretched across the north for two thousand years. In 1793, an Englishman named Sir John Barrow, who later founded the Royal Geographic Society of London, saw a section of wall near Beijing and, extrapolating from its measurements, declared that the entire structure must have contained enough stone to build two smaller walls around the equator. (Westerners rarely visited China's west, where most walls were made of tamped earth.) At that time, foreigners usually called it "the Chinese wall," but by the end of the nineteenth century, as the exaggerations accumulated, it had become the Great Wall of China. In

February of 1923, a *National Geographic* article began, "According to astronomers, the only work of man's hands which would be visible to the human eye from the moon is the Great Wall of China." (It wasn't visible from the moon in 1923, and it still isn't.)

Eventually, the misconceptions made their way back to China. Under threat of foreign domination, leaders like Sun Yat-sen and Mao Zedong realized the propaganda value of a unified barrier. *Changcheng* became the equivalent of "the Great Wall," a term that encompassed all northern fortifications, regardless of location or dynastic origin. The word described what was essentially an imaginary structure—a single, millennia-old wall.

Today, the concept of the Great Wall is so broad that it resists formal definition. When I met with scholars and preservationists in Beijing, I asked how *changcheng* should be defined, and I never heard the same thing twice. Some said that in order for a structure to qualify as part of the Great Wall it had to be at least a hundred kilometers long; others believed that any border fortification should be included. Some emphasized that it had to have been built by ethnic Chinese; others included walls built by non-Chinese tribes. Nobody could give an accurate length estimate, because there has never been a systematic survey. In 2006, various articles in *China Daily* described the length of the Great Wall as thirty-nine hundred miles, forty-five hundred miles, and thirty-one thousand miles.

There isn't a scholar at any university in the world who specializes in the Great Wall. In China, historians typically focus on political institutions, while archaeologists excavate tombs. The Great Wall fits into neither tradition, and even within a more discretely defined topic—say, the Ming wall—there's very little scholarship. The fortifications have been poorly preserved, and in the past many sections of low-lying wall were plundered for building materials, especially during the Cultural Revolution. In the 1980s, a Harvard PhD student named Arthur Waldron became interested in the relationship between Chinese and nomadic groups. "So I went to the library and thought I would find a big book in Chinese or maybe Japanese that would have everything about the Great Wall," he told me recently. "But I didn't. I thought that was strange. I began to compile a bibliography, and after a while I said, 'This does not add up to the image that we have.'"

In 1990, Waldron published *The Great Wall of China: From History to Myth*. Drawing on Ming texts—he didn't conduct significant field research—Waldron described key aspects of wall building during that dynasty. He also identified many modern misconceptions about the wall, including the notion that it's a single structure. It was a breakthrough book, and one that should have provided a foundation for further scholarship. But since then there hasn't been another work of significant new archaeological or historical research, apart from one Chinese book by a surveying team that describes a six-hundred-mile series of Ming fortifications in the east. (Another book, published in 2006, *The Great Wall of China: China Against the World, 1000 B.C.–A.D. 2000*, by Julia Lovell, a fellow at Cambridge, is primarily concerned with exploring the wall as a symbol for the Chinese worldview. She draws a parallel, for example, between the ancient wall and the current government's Internet firewalls.)

In China, one of the best-known experts, Cheng Dalin, is not an academic but a retired photographer. For more than twenty years, Cheng specialized in taking pictures of the wall for the Xinhua News Service. In his spare time, he studied history, and he has published eight books, combining photographs and research. "The Great Wall touches on so many subjects—politics, military affairs, architecture, archaeology, history," he told me. "Within each specialty, it's too small. And taken as a whole it's too big. You have to find little bits in so many different books; it's not concentrated in one place. And nobody will pay you! How will you eat? How can a person spend ten years reading all these books?"

DAVID SPINDLER FIRST STARTED HIKING THE GREAT WALL in 1994, when he was the only American studying for a master's degree in history at Peking University. He had always been athletic—at Dartmouth, he'd rowed varsity crew and was on the cross-country ski team—and he saw hiking as the perfect break from city life. At Peking University, he wrote a master's thesis in Chinese about Dong Zhongshu, a philosopher in the Western Han dynasty, in the second century B.C. After receiving his degree, Spindler decided against pursuing a career in academia. For a spell, he worked as an assistant in CNN's Beijing bureau, and then he became a China market analyst for Turner Broadcasting. But neither journalism nor business felt right, and the only constant in those years was hiking the Great Wall.

In 1997, he entered Harvard Law School. It was a homecoming—he'd grown up in Lincoln, Massachusetts—but he missed Beijing and found himself searching for distractions. ("I split a lot of wood.") During his first vacation, Spindler returned to China to hike. By then, he had the idea that in his spare time he could write a book about the Ming dynasty Great Wall, and he began reading histories. After graduation, he accepted a consulting job in the Beijing office of McKinsey & Company; every free weekend, he hiked or studied Ming texts. Finally, after a little more than a year at McKinsey, he quit in order to pursue his research full-time. His goals were ambitious: to hike every section of the Great Wall in the Beijing region and to read every word about the structure that was written during the Ming dynasty.

Spindler had paid off his law-school loans, and he had sixty thousand dollars in savings. He expected that it would take him a year or two to complete his field work. He hiked to wall sections, took notes, and recorded details on a spreadsheet. Often he saw more wall in the distance, and he marked these sitings on another database, which identified future research trips. The to-do list seemed to get longer with every journey. In 1985, a Chinese satellite survey had identified three hundred and ninety miles of wall in the Beijing region, but Spindler found many additional sections that were not included in this total.

He became a fixture in the National Library of China. He read from the Ming Veritable Records, a day-to-day history of the dynasty, and he tracked down the reports of various Ming officials. Occasionally he found a specialized work dedicated to wall defense. Some books could only be located elsewhere, so he spent weeks on the road. In a freezing library in Guangzhou, he read a detailed Ming guide to key wall fortifications; the book was so obscure that nobody had quoted it directly since 1707. He flew to Japan in order to read a rare Chinese history written by Su Zhigao, an official who served in the Ministry of Defense during the mid-sixteenth century. Spindler spent three weeks poring through the book in Japan, and during that time he ate dinner in a restaurant twice. The other nights he cooked pasta, cabbage, and tomato sauce, and poured yogurt on top. ("It's cheaper than cheese.") In Beijing, he rented a small apartment in a run-down building for $225 a month. He became familiar with the no-questions return policy of the Tilley hat. ("You have to pay for postage.") At the Miyun bus station, which is close to many sections of the wall, the minivan drivers began greeting him by shouting, "Beidianzi, six yuan!" Beidianzi is a village, and six yuan is the deal Spindler struck after an epic bargaining session that has become a part of Miyun minivan lore.

Over four years, he earned a total of $6,200 from the occasional consulting job or lecture. In 2003, he applied for a grant from the National Endowment for the Humanities, which occasionally funds projects by independent scholars. A panel of anonymous academics assessed the proposal, and they were withering. One wrote, "The applicant has no track record as an interpreter of the humanities." Another remarked: "Likelihood of completion: Not clear." The following year, with guidance from former classmates who had become professors, Spindler applied again. This time the evaluations were positive, at least in the terms of a jargon that is nearly as formalized as classical Chinese. (Panelist 1: "I feel that [the proposed book] would be a quality interpreter of the humanities.") But the application was still rejected.

In Beijing, Spindler dated a woman whom he had met when they were in law school together, and now she worked as an executive at Siemens. "She was very supportive," Spindler told me. "I couldn't have asked for anything more." But he kept finding more Great Wall and more Ming texts; finally, in 2005, they parted ways. "It's certainly one of the reasons we broke up," he said. "She couldn't see the end to it."

At the time of our first hike, Spindler had been researching for nine years, with the past four devoted entirely to the project. But he had yet to publish one word about the wall, and he had no formal contact with academia. He was extremely cautious, in part because he had grown accustomed to working in isolation. Nobody had ever combined field and textual research in this depth—it would have been impossible for any academic based in the United States or Europe—and his methodology had become as demanding and idiosyncratic as his hiking gear. In his mind, it was pointless to begin writing while his to-do spreadsheet still listed more than a hundred days of hiking.

The numbers consumed him. During our trip to the Great Eastern Tower, he commented that it was the eightieth day he'd spent on the wall that year. Since 2005, he's dated K. C. Swanson, an American freelance journalist who lives in Beijing. "David tends to remember days in relation to his wall hikes," she told me. "One day he said, 'This is our one-year anniversary.' He kept talking—he shouldn't have done that!—and he said we had started dating two days after a certain trip he took. It's like primitive cultures where people date things by when the volcano erupted."

In 2006, Spindler began giving more lectures—his main client was Abercrombie & Kent, a high-end travel service—and his income rose to $29,000. He has no hobbies to speak of, and his bookshelf is devoted almost entirely to wall research. He owns five CDs. He's surprisingly social—he's always had many close friends in the city, both Chinese and foreign, and it's common to see him out at parties in the city. But few people understand his obsession. "A couple of times I've tried to ask why it really gets him," Swanson said. "David is a very rational person, and maybe it would be easily explained in an emotional way, but that's not how David is. He's an eminently rational person doing what is basically an irrational thing."

IN OCTOBER, I ACCOMPANIED SPINDLER ON HIS 331ST JOURNEY into the field. By public bus and hired minivan, we traveled to a remote village called Shuitou. In 2003, while visiting the wall here, Spindler had seen some high ridges that he thought might contain more fortifications. In the village, he had also studied a Ming wall tablet that was now kept in a farmer's home. In 2006, the Chinese government passed the first national law protecting the Great Wall, and Ming artifacts cannot be bought or sold, but the remoteness of many sections has made enforcement difficult.

In Shuitou, we asked about the tablet, and a woman told us that the owner was out of town.

"Do you want to buy it?" she asked.

Spindler declined, and then told me, "They offered to sell it last time, too."

The harvest was nearly finished, and wind rustled stalks of corn that stood dead in the fields. Beyond the village, we climbed a steep section of wall, where thousands of Mongols had attacked in 1555. Spindler said that the typical Chinese defense relied on crude cannons, arrows, spears, and even rocks. "There were regulations about how many stones you were supposed to have, and how you were supposed to bring them to the second floor of the tower if there was an attack," he said. Later, he pointed out a circle of loose stones that had been arranged carefully atop the wall. Four and a half centuries later, they were still waiting for the next attack.

The Mongols liked to come at night. They traveled on horseback, usually in small groups. Near enemy territory, they followed ridgelines, because they feared ambushes. They were not occupiers. They penetrated Chinese lands, gathered booty, and returned home as quickly as possible. They liked to steal livestock, valuables, household goods, and Chinese people. They carried the Chinese men and women back to the steppes and allowed them to form families. Then they sent the men south to gather intelligence on Chinese defenses, using their wives and children as hostages.

The most vivid accounts of the Mongols come from Chinese officers who served in the north. Su Zhigao, the author of the rare book that Spindler read in Japan, had particularly intimate contact with Mongols during the mid-1500s. ("They like to fornicate, paying little attention to whether it's day or night or whether there's anyone watching.") Like most Ming writers, he calls them *lu*—barbarians. ("Every barbarian family brews alcohol and all of them like to drink; the barbarians drink like cattle, not even stopping to breathe in the process.") His account is a dark sort of anthropology, written in the hope that the reader will come to both know and hate the enemy. ("Barbarians like to spear babies for sport.")

In fact, they were sophisticated raiders, and warfare was complex. Both sides employed spies and spread false intelligence. Mongols signaled to each other with smoke; Chinese used blasts of gunpowder to communicate along the wall. Arthur Waldron identified three basic Ming strategies for dealing with the northerners. In the early Ming, the Chinese often took the offensive, pushing Mongol settlements away from the frontier. The second approach was buying off key Mongol leaders with gifts, official titles, or opportunities for trade. But some Ming emperors refused to negotiate with people they believed were savages. Their third option was building defensive walls—an ineffective tactic, in Waldron's view, and one that he compares to the Maginot Line. Wall building became the trademark of the later Ming, he writes, because the dynasty had become too weak to fight and too proud to conduct diplomacy.

Spindler believes that the late-Ming response was less rigid than that. In his reading, he has found that the Chinese tactics varied from place to place, depending upon local threats. And wall building was often coordinated with offensive and accommodationist strategies. In any case, he is convinced that no Chinese policy could completely resolve their problems with the Mongols, whose internal power struggles contributed to raiding. In Mongol culture, legitimate leadership was supposed to be confined to the direct heir of Genghis Khan, and to pass only to the firstborn son of each generation. Outside that narrow line, ambitious Mongols often found that the easiest opportunities to gain status lay to the south.

One such contender was Altan Khan, who was the second son of a third son. Frustrated by his genealogical standing, he attempted to improve his lot by establishing trade relations with the Chinese in the 1540s. But the reigning Ming emperor, Jiajing, refused. On September 26, 1550, the night of the midautumn festival, Altan Khan led tens of thousands of Mongols on a surprise attack northeast of Beijing. They breached the crude stone wall there and pillaged for two weeks, killing and capturing thousands of Chinese. After that, the Ming began using mortar on a large scale to improve the fortifications.

Meanwhile, the oldest son of Altan Khan, known to the Chinese as the Yellow Prince, tried a different strategy. He married dozens of women from important Mongol families as a means of solidifying alliances. But soon he began to have financial problems, which he solved in the simplest way possible: he sent the women back. The Ming had already been paying a regular quota of silver and goods to these women's families, in order to keep peace in the north; but now the ex-wives began showing up at Chinese wall garrisons, demanding more support. In 1576, after one such appeal was rejected, a raiding party penetrated a gap in a remote part of the defense network. The region was so rugged that the Ming believed that no wall was necessary, but the Mongols got through, killing twenty-one Chinese. The

Ming responded with another major wall-building campaign, this time using brick, which allowed construction on even the steepest terrain. Spindler calls the incident of 1576 "the raid of the scorned Mongol women"—a failed harem that helped inspire the stunning brick fortifications of Beijing.

Historians generally portray the Great Wall as a military failure and a waste of resources. Spindler disagrees, noting that the improved wall held back major attacks in the sixteenth century. At Shuitou, where we hiked, the Chinese defeated thousands of Mongols in a key battle. For the Ming, the wall was only part of a complex foreign policy, but because it's the most lasting physical relic, it receives disproportionate blame for their fall.

"People say, was it worth it?" Spindler said. "But I don't think that's how they thought at the time. You don't get a nation-state saying, 'We're going to give up this terrain' or 'We're going to sacrifice $x$ number of citizens and soldiers.' That's not a calculus they used. An empire is always going to try to protect itself."

IN THE AFTERNOON, WE BUSHWHACKED. ON HIS HIKES, SPINDLER sometimes followed game trails, and often he walked atop wall sections, where the brush is less dense. But occasionally there was no option other than to pursue a ridge straight through the brambles. He called this "hiking like a Mongol," and I hated it. I hated the thorns, and I hated the bad footing. I hated how my clothes got torn, and I hated the superiority of Spindler's bizarre wall regalia. I hated how branches that were chest-high for him hit me in the face. Mostly, I hated the Mongols for hiking this way.

When we reached the ruins of an old stone fort atop a ridge, it felt like emerging from a long swim underwater. To the east, the view opened for twenty miles. Only a single human settlement was visible—the village of Zhenbiancheng, still surrounded by the high stone walls of a Ming garrison. Looking down on the walled settlement, Spindler remarked that it had been a hardship post, where commanders had requested that soldiers be paid in grain rather than in silver. "There was bad inflation during the Ming," he said. "It was connected to the discovery of silver in the New World."

The next morning, after camping for the night, we discovered faint traces of another stone fort with a sight line to the ridge. Spindler theorized that they had been used to send signals to Zhenbiancheng. He carefully recorded everything, and back in Beijing he would add these details to his computer database. He was skilled at deconstructing the wall—every time I hiked with him, he noticed obscure details that reflected some element of military strategy. But eventually he would have to create something out of all these facts. For years, he had searched them out atop high ridges and in forgotten books, and sometimes the details could be as distracting as a thorn in the face. "Because David is not ensconced in academia, he's got a lot more freedom to develop his own line of inquiry," his friend Andrew Field told me. But there was a risk to the isolation. "I'm trying to urge him to seek closure," Field said. "But the way David's mind works, he has an amazing ability for detail."

The bushwhacking was intensely time-consuming, and it could also be dangerous. During our hike above Zhenbiancheng, I asked Spindler if he'd ever had any accidents. He said that in 1998, a friend had fallen off a tower and broken his wrist. They worried about hiking further through rough terrain late in the day, so they spent the night on the wall. In retrospect, Spindler admitted that he'd been too cautious, and he wished they had descended immediately.

"Was he in pain?" I asked.

"Yeah," Spindler said softly. "He was in a lot of pain."

Spindler always told friends where he was hiking, and he almost never made overnight trips alone. His most frequent companion is Li Jian, a classmate from Peking University who now works in the rare-books division of the National Library. Her first expedition with Spindler, in 2000, was a three-day hike. "I had always had problems with insomnia," she told me. "But when I got back from that hike I slept really well!" Since then, she has spent 185 days on the wall with Spindler.

Over time, Li Jian acquired an L.L.Bean wool hunting shirt, a white Tilley hat, La Sportiva mountaineering boots, and elk-leather utility-line worker gloves from J. Edwards of Chicago. She cut off a pair of long underwear at the knee and scissored a round hole for her face. In the field, she's a five-foot-two female Chinese double of Spindler, following him through the brush. She told me that she has never led.

In June of 2003, they set off for a three-day hike in the wilds of Mentougou, in the western Beijing region. The mountains there are weirdly shaped; the high peaks are easy to negotiate, but the lower flanks deteriorate into unexpected cliffs. Hiking like a Mongol, Spindler got lost, and every attempt to go down ended at a sheer drop. Fortunately, it had rained, so they drank water that had settled in the hollows of rocks. Friends organized a search party and drove out from Beijing. Five days after Spindler and Li Jian had set off, they finally found a trail and made it back, meeting the search party en route. Today, Li Jian continues to hike the wall as therapy for insomnia.

CHINESE UNIVERSITIES MAY NOT HAVE PRODUCED GREAT WALL specialists, but a small community of wall enthusiasts has developed outside of academia. They tend to be athletic—a rare quality among the Chinese intelligentsia, whose disdain for labor has sometimes been a problem in archaeology. And the Great Wall attracts obsessives. Dong Yaohui, a former utility-line worker, left his job in 1984 and doggedly followed wall sections on foot for thousands of miles across China. After writing a book about the experience, he helped found the Great Wall Society of China, which publishes two journals and advocates preservation of the ruins. Cheng Dalin, the retired Xinhua photographer, graduated from a sports academy. William Lindesay, a British geologist and marathoner, came to China in 1987 and ran and hiked more than 2,470 miles along the wall. He settled in Beijing, published four wall-related books, and founded International Friends of the Great Wall, a small organization that also focuses on conservation.

The most active Great Wall researcher at Peking University is a policeman named Hong Feng. As a child, he enrolled in a sports school—he became a sprinter and a long jumper—but he always enjoyed reading history. After barely missing the cutoff for admission to college, he entered the police academy, and was eventually assigned to the city's unit at Peking University. In the mid-1990s, he began hiking recreationally and was disappointed with contemporary books about the wall. "They make too many mistakes," he told me. "So I started reading the original texts."

I met Hong Feng in the Peking University police station, where he was working a twenty-four-hour shift. He is the station's supervisor, and uses his days off for hiking trips. At forty-five, Hong is tall and strong, although he suffers from a chronically sore right elbow, which was injured when he fell while researching. He often visits the university library, but he has never discussed his research with professors. "Scholars in the archaeology and history departments just aren't interested in the Great Wall," he said.

During his hikes, Hong noticed a puzzling fifteen-mile gap in fortifications to the northwest of Beijing. Modern writers had claimed that the landscape was so rugged that it didn't require defenses, which made no sense to Hong. He had visited other areas that were much steeper yet heavily fortified, so he turned to the Ming Veritable Records. He discovered that the Ming believed the region contained an important *longmai*, or dragon vein, just north of their ancestral tombs. A dragon vein is a ridgeline critical to feng shui, so the Ming went to the trouble of building elaborate walls farther north, on terrain that was naturally less defensible.

Hong Feng published an article about his findings on www.thegreatwall.com.cn, which has become home to the most vibrant community of Chinese wall enthusiasts. The site was launched by Zhang Jun, a software engineer, on May 8, 1999—the day the Chinese embassy in Belgrade was bombed by NATO. (NATO said that the attack was a mistake.) Members of the Web site have regular dinners in Beijing, and at one of the events I asked Zhang Jun why he had been inspired to found the site on that particular day. "You can say that the Great Wall was built to protect China," he said, choosing his words carefully.

The Web site has five thousand members, many of whom are interested in the wall for a combination of patriotic and recreational reasons, although there's also a small community of serious researchers. David Spindler joined in 2000. Like everybody else, he adopted an online name—Spindler's is Ah Lun, a derivative of the Chinese name that he was given by a language teacher—and he frequently corresponds with others in Chinese. But he doesn't attend functions, and he's never identified himself as a foreigner. Starting last fall, he posted two long articles on the Web site, describing the construction history of specific sections of wall. He told me that he would eventually write his book in English, but for initial articles it made sense to write in Chinese, because the Web site is the only community that cares about such esoteric topics. (One of his articles was titled "On the Construction Date of the 'Pig's Mouth Fort' Great Wall.")

Spindler had asked me not to identify him to the other members of the Great Wall Web site, and I didn't, but they quickly brought up Ah Lun on their own. Hong Feng, the policeman, spoke admiringly of Ah Lun's research, assuming that he was Chinese. "He doesn't write very much, but what he writes is deep," Hong said. "He must be some kind of graduate student or scholar. I don't ask, and he doesn't tell."

Eventually, Spindler planned to "come out" as a foreigner, but he had always been wary of the site's nationalism. And he remembered the way he felt after defending his thesis at Peking University. "My professor said, 'In the rules for foreigners, we usually give them a little more latitude,'" Spindler told me. "If I had had more presence of mind, I would have said, 'Well, I've been here for the experience, and I'll be happy to walk away without a degree.'" He continued, "I want my work to be evaluated on these stand-alone terms. Who it's written by, whether he's Chinese or foreign, shouldn't matter."

It seemed contradictory—Spindler published under a pseudonym because he was worried that he wouldn't get the credit he deserved for his work. But this was characteristic of many of his actions. He was extremely cautious, but somehow he had risked everything—financial stability, relationships, personal safety—for his research. He had full confidence in his ideas about the wall, and he described them with perfect clarity; it was obvious that he wanted to help dismiss all the popular misconceptions about the structure. And yet he refused to start writing his book before the spreadsheets satisfied him. His work could be seen as a study in obsessions—a man's single-minded pursuit of one of the most ambitious structures on earth. But beneath it all lay a deep commitment to rationality. Spindler believed that the wall had been built for a military reason, and he believed that he was researching it in the best way possible. He hated any symbolic use of the Great Wall, especially for something as complex as Chinese culture. For Chinese, the wall usually represents national glory, whereas foreigners often see it as evidence of xenophobia. But Spindler felt that neither interpretation was useful. "It's just one manifestation of what China has done," he said. "It's just a way they defended themselves."

Of all the people I met, Hong Feng had a viewpoint that reminded me the most of Spindler's. Hong's online name was Qiong Shishu, which means "to reach the end of the books."

"People in China always describe the Great Wall as a symbol of ethnic pride," Hong told me. "But that's an exaggeration. It wasn't supposed to be a great monument like the Pyramids. It was built in response to attacks."

AT THE END OF DECEMBER, I ACCOMPANIED SPINDLER ON HIS 340th trip along the wall. During a previous visit to Miyun, north of Beijing, he'd seen some high ridges that he believed might contain towers of piled stone. Slowly, we climbed to the ridges: Nothing. But it was another day to be checked off on the to-do list.

Although I had never liked the bushwhacking, during the past year I had come to appreciate the distinctive rhythm of the trips. Every journey had it all: good trails, bad trails, hellish thorns, spectacular views. No matter the landscape, I could always see Spindler up ahead, moving steadily, his Tilley hat bobbing above the thickets.

On the way down, we found a dead roe deer in a trap. The loop snare had caught the animal around the neck; it must have strangled itself. Just beyond that, we reached a long section of wall where most ramparts had crumbled away. As I walked atop the structure, my boot got caught in a hole. I tripped and fell down a short ledge, pitching headfirst toward a ten-foot drop. Somehow—things happened very fast—I threw myself down against the wall. I slammed to a stop with my head peering over the edge.

"Nice save," Spindler said, after he had rushed over. I rose slowly, and tried to walk, and knew that my left knee was badly hurt. But we were miles from help, and the temperature was well below freezing; the only option was to keep moving.

During the descent, I leaned on Spindler whenever possible. It took three hours, and I remember every minute. The next morning I went to the hospital for X-rays. The doctor told me that I'd broken the kneecap in multiple places, and I'd be on crutches for six weeks; and that was the last time I walked on the Great Wall of China.

THE DAY AFTER THE ACCIDENT, SPINDLER STOPPED BY MY apartment. He asked if there was anything I needed, and I could tell that he felt bad about what had happened. He mentioned that he had made a quick analysis of the spreadsheets, which showed that mine was only the second casualty to be sustained in approximately 1,250 person-days of hiking. Later, he confirmed that the exact figure was 1,245.

In February, before leaving on a research trip to Taiwan, he visited me again. He planned to study some Ming maps and memorials that were held in Taipei's National Palace Museum. He still hadn't written anything in English, but additional Chinese articles were in the works, and he seemed to be thinking more about the future. He planned to start writing the book within a year or so; after it was finished, he'd find a way to continue researching the wall in other parts of China. Maybe he'd start a PhD program, or perhaps he'd remain independent, supporting himself with lectures and books. "I'll need to learn other languages in order to get academics to give me the time of day," he said. "You really get written off if you don't know Japanese, if you don't know Mongolian. There are others that would be helpful. Probably the next one would be Russian, and possibly German. I guess it would be helpful to learn some Manchurian. A little Tibetan. But those are further down the list."

I hopped on crutches to the door and said good-bye. He had an early flight; in order to save money, he'd booked a ticket with a seven-hour layover in Macau, where he'd be restricted to the airport. He figured he'd do some reading. When I asked how many Beijing hiking days were left, he didn't hesitate. "Eighty-six," he said.

# THE DIRTY GAME

JAY-Z LOVES THE PEACE CORPS. HE'S NEVER SAID SO PUBLICLY, and there's no reference to volunteerism in any of his 224 songs, not even the one titled "Money, Cash, Hoes." But Rajeev Goyal believes that he knows the rapper's true heart. "Jay-Z and Beyoncé are both very interested in helping the Peace Corps," Rajeev told me once. He said that last year he was on the phone with somebody who claimed he could arrange for Jay-Z and Beyoncé to speak at a Peace Corps rally that Rajeev was organizing in Washington, D.C. But their appearance fell through, which sometimes happens to Rajeev's most ambitious plans. He failed to get a Grateful Dead guitarist to telephone a Vermont senator. He was unable to get an audience with the Dalai Lama in Dharamsala to request a letter from His Holiness asking Congress to give more money to the Peace Corps. Once, he asked Maureen Orth, a writer for *Vanity Fair* and Tim Russert's widow, to contact Senator Barbara Mikulski, of Maryland, in a manner so roundabout that it was like driving from D.C. to Baltimore via the Deep South. "He asked me to ask James Carville to ask Bill Clinton to call Senator Mikulski," Orth told me. "And that's just one of four e-mails that I got from him in a day!" Orth didn't telephone Carville, but on another occasion she called a senator on his cell phone in the middle of a meeting.

"It was outrageous, but I did it for Rajeev," she said—like everybody, she used his first name when talking about him. Orth admired Rajeev's willingness to try anything, especially since he had appeared in Washington as if "he was dropped in there from a cloud." She said, "Who else would fly on miles all the way to Hawaii to try to see Obama's sister? And get it done! I wish he had been a reality series."

Rajeev Goyal is thirty-one years old, but he could pass for a college student. He stands only five and a half feet tall, with dark skin and long-lashed eyes. He has the portable confidence of the second-generation immigrant—no matter where he goes, he knows there are benefits to being an outsider. In the part of eastern Nepal where Rajeev served as a Peace Corps volunteer from 2001 to 2003, people sometimes weep when his name is mentioned. Locals refer to him as Shiva, the god who is also the source of the Ganges River. Old folks turn on a tap and say: "This is what he gave us." In the halls of Congress, most people have no idea what to make of him. For the past two years, he's approached the place as if it were just another Nepali settlement with a caste system to untangle. He figured out the Washington equivalent of village well routes—hallways, hearing rooms, and coffee shops where anybody can hang around and meet a member of Congress. "He just picked off Democrats and Republicans one by one," Sam Farr, a Democratic congressman from California, told me. "I don't know lobbyists who are that persistent." Others complained that his unorthodox approach was too personal, but even critics acknowledged the results. During the past two years, funding for the Peace Corps has increased by record amounts, despite partisanship in Congress and a brutal economic climate. "I've been in the Congress for seventeen years, and always lobbying for the Peace Corps, but I've never been as effective as I have in the last two sessions," Farr said. "And I would attribute that to Rajeev."

In March of 2011, the Peace Corps will turn fifty years old.

The anniversary is bittersweet: despite the new funding, which has allowed for a significant increase in volunteers, the agency sends fewer than 60 percent as many people abroad as it did in 1966. Many Americans aren't aware that the Peace Corps still exists. Its impact on foreign policy seems minimal, especially in light of the recent wars in Afghanistan and Iraq. Rajeev told me that the agency might have a higher profile if former volunteers applied lessons from the developing world to U.S. politics, which is the opposite of how most people think of the Peace Corps. Instead of introducing American values to some benighted country, Rajeev wants to move in the other direction. "The way we organized this campaign was asking who was in the Peace Corps, and who cares about it. That's your blood link; that's your caste. You define your tribe." He said, "Washington is a village. Decisions in Congress, some of the biggest decisions, are based on a personal act."

RAJEEV GOYAL DIDN'T KNOW HIS OWN CASTE UNTIL HE JOINED the Peace Corps. He grew up in Manhasset Hills, Long Island, where his parents, Ravindra and Damyanti Goyal, had settled after immigrating from Rajasthan, India, in the early 1970s. They raised their three sons to speak Hindi, but they never told them they were Vaishya, a caste known for its success in business. "To us, everybody is equal," Ravindra Goyal, who is a cardiologist, told me, explaining that he didn't like the caste system. But after the Peace Corps sent Rajeev to Nepal, which has a system similar to India's, he telephoned his mother. "He asked me, 'Are we Brahmin, or Vaishya, or Kshatriya, or what?' Damyanti told me. "He said, 'People want to know what I am.' So I told him."

The Goyals had initially opposed their son's decision to go abroad. They wanted him to become a doctor, and he took pre-med courses at Brown University before deciding to apply to law school. They didn't see the point of deferring admission in order to live in an undeveloped country. Damyanti, who had recently undergone chemotherapy for breast cancer, worried about her son being far away, but his arguments finally swayed her. "He said that this country has given us too much," she told me. "We have a nice house, nice car, we live in a nice neighborhood. It's time to give something back. When he explained it like that, I liked the idea a little better."

Rajeev was assigned to teach English at a school in Namje, a village of fewer than six hundred people. Namje is located in eastern Nepal, where the Terai Plain gives way suddenly to the foothills of the Himalayas. At an elevation of more than five thousand feet, Namje villagers grew coriander, soybeans, radishes, and other vegetables. They also spent much of their time hauling water. Snowcaps provide Nepal with abundant water resources, but rivers are often inaccessible in mountain towns like Namje. The nearest source was the Saacho Khola, a spring that was more than two hours away by foot on steep mountain paths. People often made three trips a day, carrying sixteen-liter aluminum jugs. "You learn that not everything needs to be washed," Rajeev told me. "Soap isn't always necessary. You waste a lot of water with soap. Of course, I didn't do this as well as they did, so I got sick."

He caught a case of scabies so bad that it scarred his arm. After visiting a doctor in Kathmandu, he returned to the village and noticed other effects of the water shortage. "One day a good student didn't come to class, and I asked him why. He said he was getting water. I brought all the villagers together and asked, 'If there's a way to solve this problem, are you willing to donate your labor?' They were willing to do it."

Rajeev spoke with engineers in Kathmandu, and he read books about electric pumps, piping, and filtration systems. Using the skills he had acquired in premed physics classes, he learned to calculate water friction. He finally decided that the best option was a two-stage pumping system capable of lifting water thirteen hundred vertical feet. In the city of Dharan, he found a pipe salesman named Kishan Agrawal, whose ancestors came from the same part of India as Rajeev's. After the two men discussed their family histories, Kishan agreed to order hundreds of pieces of three-inch galvanized iron pipe on credit, interest-free. In order to raise funds, Rajeev returned to Long Island, where families in the Indian immigrant community often host big weekend meals that include singing and dancing. The Goyals arranged a dinner and invited their doctor friends, without mentioning that the night's entertainment would involve Rajeev asking for money. "I had the idea to do the fund-raiser here," Damyanti told me. "I knew that if we didn't have a fund-raiser, we'd end up paying for all of it ourselves." Within ten minutes, Rajeev had raised $18,000, which was eventually supplemented by funds from the Peace Corps, USAID, and the American Himalayan Foundation.

For the twenty-two-year-old Rajeev, the most daunting challenge was organizing labor. The villagers had no power tools, and all materials had to be carried via mountain paths to construction sites. Women did much of the work, because many men had gone abroad as laborers. Through his research, Rajeev learned that one of the biggest threats was something called water hammer—the pressure that builds in a long pipe when the lower outlet is closed. He consulted with engineers, who suggested building a stone staircase more than a mile in length, which would pin down the pipe and also allow access for repairs. In designing the staircase and the pump houses, Rajeev relied heavily on Karna Magar, a villager who was naturally gifted but had only a ninth-grade education. Harka Lama, the headmaster of the local school, organized the village into twenty-five groups, each of which would help coordinate different aspects of the project.

Another teacher named Tanka Bhujel handled village politics. "I'm only realizing now how much he taught me," Rajeev told me recently. "We would go to a meeting and he would say, 'If we can get this one guy, we'll get everybody.' And the guy would have three wives and things would be complicated. So much was based on ancestry and bloodlines. It's the same in Washington. It's identity." Tanka was an outsider in the village, a member of an obscure subcaste, and he relished the political maneuvering. "Tanka would be speaking in Nepali to a group, and then out of nowhere he would say in English, 'Po-litics is the dur-ty game,'" Rajeev said, mimicking a Nepali accent. "And he wouldn't translate it! He understood Kathmandu, he understood the Maoists, the military, the family politics."

At the time, Nepal was ripped apart by unrest. Since the mid-1990s, Maoist groups had been trying to overthrow the monarchy, and after peace talks failed in 2002 there were increased attacks on the army. Often, the military responded with brutal violence as soldiers searched villages. Thousands of people were killed, and more than a hundred thousand were displaced. One of the main guerilla tactics involved setting off homemade bombs, which meant that anybody with a pipe was suspect. Rajeev got the local army commander to issue a letter explaining why villagers were handling so much plumbing equipment, and everybody carried a copy at all times.

Five hundred and thirty-five people volunteered for the project. Entirely by hand, they built two pump houses, two holding tanks, three reserve tanks, and 1,236 stone steps—said to be the longest staircase in eastern Nepal. When an agent from the district engineering office visited, he was shocked that an American like Rajeev would depend so heavily on the uneducated Karna Magar. "Rajeev Sir, you're from a developed country, and you don't have an engineer on this project?" the agent said in disbelief. Rajeev answered, "Our guy is as good as any engineer." But after sixteen months of work, they turned on the power and nothing happened.

Rajeev had only a month left in the Peace Corps. After repeatedly failing to get the pump to work, he became so discouraged that he hardly listened when a local electrician suggested the problem might be voltage. "All those bastards in India, they're using too much electricity," the electrician said. "We have to wait until the middle of the night, when all those people in India will be asleep."

A group of men hiked to the lowest pump station, where they fished for salamanders while waiting for India to go to bed. Rajeev was so depressed that he stayed home. At three o'clock in the morning, a neighbor woke him up. "*Pani aayo!*" he shouted. "The water has come!" Rajeev ran to the staircase, where he heard a sound like rain: water was rising in the pipes. He and others followed the noise up the mountain, step by step. It became louder at the summit, as water poured into the series of reserve tanks, which held almost twenty thousand gallons. Tanka Bhujel, who knew the politically correct response to any occasion, went out and slaughtered a goat.

PEOPLE USUALLY ASSUME THAT THE PEACE CORPS CAME OUT OF A grand idea, but its beginnings may have had more to do with emotions associated with village politics. In October 1960, during the third debate of the presidential campaign, Richard Nixon attacked John F. Kennedy by claiming that Democratic presidents had been responsible for leading Americans into every war of the past half century. Immediately afterward, Kennedy flew to Michigan, where, at two o'clock in the morning, he made an unplanned speech on the Ann Arbor campus. He challenged the students: "How many of you who are going to be doctors are willing to spend your days in Ghana?" As Stanley Meisler describes in *When the World Calls*, Harris Wofford, an aide who later became a senator, believes that Kennedy made the speech because he was angry about Nixon's insinuation.

Thousands of students sent letters of interest to Kennedy, and the idea also tapped into the popular feeling that the United

States needed more grassroots efforts to fight communism. After the election, Kennedy founded the Peace Corps and appointed his brother-in-law, Sargent Shriver, as director. Shriver moved fast—in less than six months, the first volunteers were sent to Ghana, where they had an immediate impact on the education system. Without the Peace Corps, roughly a sixth of Ghanaian secondary schools would have had to shut down owing to lack of teachers. Soon there were nearly five hundred instructors in Ethiopia alone. By 1966, more than fifteen thousand volunteers were working in various jobs around the world, and the agency received forty-two thousand applications that year. Plans for expansion were ambitious: Kennedy once remarked that the Peace Corps would start to become significant when it reached a hundred thousand volunteers.

But 1966 turned out to be the high-water mark. Applications plummeted during the Vietnam years, when idealistic young people weren't inclined to have anything to do with a government agency. During the 1970s and 1980s, volunteers continued to go overseas, where they often had a major impact on communities, but the agency's U.S. profile diminished. Support depended on the whim of a president or a few politicians. Nixon tried to kill the Peace Corps entirely—he hated anything associated with Kennedy. Carter appointed a director who turned out to be incompetent. Reagan was surprisingly supportive, especially after a 1983 meeting with the prime minister of Fiji, who effusively praised the volunteers who had served in his country. A week after that meeting, a staffer presented Reagan with a proposal for slashing the federal budget. "Don't cut the Peace Corps," Reagan reportedly said. "It's the only thing I got thanked for last week."

Over time, though, the Peace Corps came to embody the empty campaign promise. Everybody had heard of it, and impressions were vaguely positive, but there was no real awareness of what volunteers did or how their activities were funded.

Clinton claimed that he would increase the size of the organization from fewer than seven thousand volunteers to ten thousand; George W. Bush said he wanted fifteen thousand. Obama promised to double the Peace Corps by its fiftieth anniversary. But none of them pushed hard for more money, and volunteer numbers stayed at roughly half the level of 1966, despite the fact that applications increased significantly after 9/11. In 2008, the Peace Corps budget was $342 million—less than what the federal government spent on military bands.

To former volunteers, it seemed a wasted resource. The Peace Corps had sent Americans to Afghanistan for seventeen years, and more than forty-five thousand people had served in predominantly Muslim countries, but these things seemed to have no effect on post-9/11 policy. Kevin Quigley, the president of the National Peace Corps Association, a group for returned volunteers, believed it was time for a campaign to expand the organization. But this had to be done independently of the Peace Corps—by law, a government agency can't lobby. Quigley told me that the community of former volunteers had been too passive. "You have to get organized," he said. "You need a mechanism so legislators know that this is a significant issue."

Quigley met with Donald Ross, a former volunteer in Nigeria who had organized public campaigns for Ralph Nader and others. With a grant from the Rockefeller Brothers Fund, they hired Rajeev Goyal, who, since leaving Nepal, had attended law school at New York University. Rajeev had continued doing development work in Namje, but he had never had any contact with Capitol Hill. At first, he tried to read up on the legislative process. "They have those books, *How Our Government Works*, whatever," Rajeev told me. "It's the most useless waste of time." He realized that there's no legal or democratic element—agencies like the Peace Corps are funded through appropriations committees, which aren't outlined in the Constitution. All that matters is the personal decisions of committee members, and how they can be influenced by constituents, colleagues, and other people.

The grassroots part was relatively easy. There are more than two hundred thousand former volunteers, and Rajeev eventually developed an e-mail list of thirty-three thousand. He installed computer software that detected whether a message had been opened, which taught him what kind of e-mail inspired people to read and forward. He also tracked his targets. Once, when I visited him in Washington, he checked his computer and told me that a recent message had been opened 133 times by staffers in the office of a certain senator. "That means I did something right," he said. With his list, he generated enormous numbers of phone calls and e-mails from former volunteers across the country. During the week of one key funding decision, so many people called the office of Representative Nita Lowey, the chair of the House Appropriations Subcommittee on State and Foreign Operations, that Lowey's phones were completely tied up. An aide finally begged an intermediary to convince Rajeev to stop. "Call off the dogs," the intermediary said. "The interns need to go to lunch."

Meanwhile, Rajeev contacted the kind of influential people who are known as "grasstops." He looked for Peace Corps connections: the MSNBC anchor Chris Matthews had been a volunteer in Swaziland; the chairman of the Chicago Bears served in Ethiopia; Jimmy Carter's mother had been in India, and his grandson in South Africa. When it came to asking for favors, Rajeev was fearless—he once tried to get all four living former presidents to sign a letter to Obama, inspiring Jimmy Carter to write back, "This is no way to approach a former President of the United States." But Carter made a call on behalf of the campaign. (I first met Rajeev when he asked me to write a letter about my years in the Peace Corps in China.) After Rajeev heard that Obama's half-sister once considered joining the Peace Corps, he solicited airline miles from another former volunteer, flew to Honolulu, put on a Hawaiian shirt, and met with Maya Soetoro-Ng. By the time he left Hawaii, he had one letter from her and another from Obama's former high school teachers. For good measure, he also stopped by the president's favorite shaved-ice stand and got a pro–Peace Corps message from the proprietor.

Maya Soetoro-Ng told me that she had never endorsed another campaign of this sort. But as an educator who had grown up in Indonesia, she supported overseas service, and Rajeev impressed her. "He seemed very professional, very affable," she told me. "But he wasn't particularly slick about it. He was very natural and pleasant." She emphasized that she didn't represent her brother or his administration, which until that time had issued no public statements about the Peace Corps. But a week after Rajeev's Hawaii trip, Michelle Obama gave a speech in which she said, "My husband is committed to substantially expanding the number of opportunities to serve in the Peace Corps."

In Namje, Tanka Bhujel had taught Rajeev that individuals often matter more than the system. "His style is to go directly to the most powerful person and ask what he wants," Rajeev told me. Initially, he found this hard to do in Washington, where somebody without connections can't get on the schedule for an important member of Congress. But then Rajeev started studying a book with color photographs of everybody in the House and Senate. "Once you know somebody's name, you can talk to them," he explained. Over time, faces stuck in his memory, and that was when Washington truly became a village. He recognized Senator Bob Corker and Senator Christopher Dodd at Reagan National Airport. He struck up a conversation with Representative Russ Carnahan at the Starbucks on Pennsylvania Avenue. He ran into Representative Peter Welch late one night outside Cosí, and he met Representative Dennis Kucinich at Le Pain Quotidien. Anybody can wander around the Senate hallways—you don't even need an ID to get in—and Rajeev spent days there. He attended committee hearings so that he could approach key officials during breaks. He learned that the best spot in the House is the small underground rotunda that connects the Cannon and Longworth buildings; he met dozens of members there. His second year on the job, 124 House members added their names to a "Dear colleague" letter in support of increased funding for the Peace Corps, which drew more signatures than any other issue.

People told him that this routine was called "bird-dogging." Lobbyists rarely work like this, because elected officials don't want to be seen in public with a special interest, but the Peace Corps is far less threatening, especially when represented by Rajeev. "It's rare that they see somebody a little bit young and a little bit brown," he said. In August, I spent time with him on the Hill, and within two and a half days he had talked to fifteen senators without appointments. It was as if the political world had suddenly become very small, and yet it had a distinctly exotic tilt—Rajeev could find a Nepali connection that would start almost any conversation. He approached Mark Udall, the senator from Colorado, by mentioning that Udall's mother had also served as a volunteer in Nepal. The senator immediately brought his hands together, bowed slightly, and said, "*Namaste*." Rajeev introduced himself to Senator Dianne Feinstein by noting that her husband was the honorary consul general to Nepal. At a breakfast for Iowa constituents, Rajeev caught Senator Tom Harkin's attention by referring to another Midwesterner who exports Tibetan rugs from Kathmandu. When he mentioned the Peace Corps, the senator said, "I wonder how many volunteers we have in Haiti."

"None," Rajeev said. "Zero."

"None?" Harkin said. "That's unconscionable!"

"There's no money," Rajeev said. "They cut the Haiti program a number of years ago, because of unrest, but now they need to bring it back. That's why we're asking for an increase.

I know that you were in Vietnam recently, and they'd like volunteers as well."

"I just got back from Vietnam," the senator said. "There's no
Peace Corps there?" He called to his foreign affairs aide, who
was across the room. "Tom! Tom! Agent Orange and the Peace
Corps!" The aide hustled over, no doubt wondering what the
short Indian had to do with Agent Orange and the Peace Corps.
"That's what I want you to check on," the senator explained,
mentioning Patrick Leahy, the chairman of the Senate Appropriations Subcommittee on State and Foreign Operations, which
funds the organization. "Talk to Leahy about the Peace Corps
and Agent Orange."

When I was with Rajeev, he attended a public confirmation
for the ambassador to East Timor, an armed forces hearing about
Russian nuclear weapons, and a meeting dedicated to rare and
neglected pediatric diseases. To him, this was all background
noise; he sat in the back and wrote e-mails to constituents on his
iPhone while waiting to bird-dog during breaks. Between sessions, he called Jimmy Carter's grandson and the daughter of
Daniel Patrick Moynihan. He studied the proposed budget that
had come out of the House, making sure he knew the figures.
Certain facts were always on the tip of his tongue: the entire
Peace Corps budget is less than the price of two F-22 jet planes.
He calculated that it costs approximately $25,000 a year to send
a volunteer abroad, versus a million for a Foreign Service officer.
The proposed budget for 2011 allocates twelve times as much to
the militaries of foreign countries as it does to the Peace Corps.
When he approached Senator Feinstein, she explained that a big
increase would be hard in this economic climate.

Rajeev took a copy of the budget out of his pocket. "Look at
the foreign military finance," he said. "It's almost five and a half
billion dollars, and that includes a one-point-two-billion-dollar
increase. All we're asking for is dust compared to that. It's forty-
six million."

The senator raised her eyebrows. "Just dust?" she said.

"That's it."

"Well, we'll take a look at it."

"Look at this money," Rajeev continued. "A one-point-two-billion-dollar increase for foreign militaries, and nine hundred million for Pakistan counterinsurgency."

Feinstein had turned to go, but the numbers caught her attention. "How much is that?"

Rajeev held out the paper. "One point two billion and nine hundred million."

"Let me take a look at that," she said, and he handed over the paper. She called to an aide: "Give this to Rich."

Rajeev told me that he sometimes gets into trouble for being too bold. He had recently got kicked out of a military reception for Rhode Island veterans, where he had hoped to meet a senator. When asked who he was, Rajeev responded that he was a Brown graduate and "a Peace Corps veteran." "To succeed you have to be a little bit cunning," he told me. He didn't believe there was necessarily anything dirty about politics, but he agreed with Tanka Bhujel's description of it as a game. "It's part of being human—it's a human game," Rajeev said. "Like all games you should relish it and take pleasure in it. People believe that in order to be in politics you have to be a politician. That's not true at all. You can be the lowest villager and still be involved."

RAJEEV OCCASIONALLY GOT IN TROUBLE IN NAMJE, TOO. EVEN before the water line was finished, he knew that there would be political complications, especially after workers discovered that somebody had taken a shit in the lower holding tank. That was the simplest political statement: it meant that a neighboring village was unhappy. The Namje residents called a meeting and eventually agreed to give their neighbors a ten-thousand-dollar water line. While Rajeev was at law school, he raised funds, and over the years the villagers steadily extended the project, until it served seven small communities. Water fees allowed Namje to hire three full-time maintenance workers, and they used the extra profits to add another salaried teacher at the local school.

Rajeev also worried about what Namje women would do now that they no longer spent six hours a day hauling water. Before he left the Peace Corps, he founded a women's co-op, although he had no clear idea of its future activities. As he was leaving, a woman presented him with a gift of a traditional hand-knit woolen cap. The craftsmanship impressed Rajeev, who said, "I'm sure you could sell this for fifteen dollars in New York City."

The first box of hats arrived as he was beginning law school. It was more than four feet tall, and it had been shipped to the home of Ravindra and Damyanti Goyal, on Long Island. It didn't take long for the second box to appear. The boxes had not been well packed and they smelled like wool that's wilted a little after the Himalayan monsoon. The Goyals requested that future shipments be directed to the West Village, where Rajeev was living with his older brother Rishi, who had begun his residency at Columbia University Medical Center. The apartment measured eight hundred square feet and soon the living room was devoted entirely to boxes.

On weekends, Rajeev stood on Houston Street and sold hats. It was near the law school, and he often saw fellow students and professors, who said, "Rajeev, what the hell are you doing?" Rishi told me that he didn't like having an apartment full of rotting hats, but there was some consolation in watching his brother work as a vendor. "He wouldn't get a permit," Rishi said. "He brings a table from God knows where, and he gets into squabbles with the other people who were supposed to be there. He enjoys that little squabble. He thinks that's the stuff of life. It was a good show. I'd get a cup of coffee. He'd be standing there in the freezing cold. He would target couples and play the liberal guilt."

Back in Namje, it became impossible to buy a hat, because prices were inflated to Manhattan levels. The women's co-op swelled in membership as people drifted in from other villages to make hats. Finally, after Rajeev had sold five thousand dollars' worth—he was also hawking them in the back of tax law class at NYU—Tanka Bhujel put a stop to it. He called a village meeting and said, "This is unsustainable." Five years later, when I visited the Goyals in Long Island, they still had more than six hundred Nepali hats.

Rajeev did not like law school. He found most classes boring, and he got the only D that was handed out in criminal procedure. But he graduated, and he passed the New York bar exam easily. Still, much of his energy went into organizing fundraisers, and he returned regularly to Namje. By 2004, the Maoist conflict was so bad that all Peace Corps volunteers were evacuated from Nepal, but Rajeev started a series of school-building projects. Some local communities had terrible facilities, and the Namje crew could build a seven-room school for about $25,000. They always applied Harka Lama's strategy of dividing villages into volunteer work groups—other Nepali NGOs have since adopted this method. Karna Magar, now nicknamed "Local Engineer," helped with technical issues. Priyanka Bista, a Nepali-Canadian architecture student, designed two of the schools with him. Through these projects she met Rajeev, and they fell in love, marrying in Queens, where priests from both India and Nepal officiated.

Kishan Agrawal, the pipe salesman whose family had come from the same region as Rajeev's, became more active in the projects. Kishan was a Rotary Club member in the city of Dharan, and one of Rajeev's uncles was a Rotarian in Plainsboro, New Jersey. Somebody in the New Jersey club suggested that they raise funds and have them matched by Rotary grants. The only problem was that Rotary International has a policy against using matching grant funds for construction, in part because of legal and liability concerns. But the Rotarians assured Rajeev that this wouldn't be a problem; they could build facilities and simply describe them as "school supplies." Nobody would ever know the difference unless Rotary International sent somebody all the way to the Himalayas to audit the project. Rajeev raised $28,000 at NYU, and Rotary contributed another $20,000 in grants, and two schools were built. Tanka Bhujel bought $40 worth of school uniforms, pens, and notebooks, took photos of smiling children with the gear, and mailed them off with letters certifying that the project had been for supplies.

Rotary International announced that it was sending a member from South India to audit the project. On a hot day, he arrived wearing a button-down shirt, a necktie, and a white loincloth. When I asked Kishan what the auditor looked like, he said, "He was very healthy," which is the Nepali way of saying "massively obese." In the town of Karkichap, the Rotarian became emotional when children poured out of the beautiful new school building and adorned him with ceremonial flower necklaces. He tried to continue on foot to the second Rotary-funded school, but after five minutes he was overwhelmed by healthiness and had to sit down. The teachers from the other school hiked up to visit him there instead. "You've done a great thing," Kishan remembers him saying.

"So everything is fine?" Kishan asked.

"No, I'm afraid you're in big trouble," the Rotarian said. "You're not supposed to build schools."

For the next year, Rajeev engaged in evasive correspondence with Rotary International, which wanted its money back. His uncle stopped attending Rotary meetings in Plainsboro, which along with the Dharan chapter was suspended from Rotary grant programs until the dispute was settled. Finally, Rajeev and Tanka decided to blame everything on the Maoists. They wrote letters claiming that they had fully intended to purchase school supplies, but Maoists came to the village and forced them to build the schools instead. Rajeev dipped into his law school loans, and he and his uncle and various Rotarians sent a total of seven thousand dollars to Rotary International, which finally resolved the matter. "It completely wiped me out," Rajeev said.

When I met Kishan in Nepal, he told me proudly that he no longer attends Rotary meetings. "I didn't go according to the Rotary system," he said. "I went according to Rajeev's system." He said that even during terrible times Namje and nearby communities didn't become violent, because people were busy with development work. And Kishan said that Rajeev had changed his life. "I realized that every person should be involved," he said. "You need to do something for other people."

I stopped by the school in Karkichap, which looked deceptively peaceful, considering that it had destroyed Rotary careers all the way from Dharan to Plainsboro. On the bright-blue roof, somebody had painted an enormous white Rotary symbol that could be admired from any low-flying aircraft. The headmaster told me that Nepali and Indian Rotarians sometimes travel long distances to see the building. The library had a beautiful door made of sisau wood, more valuable than mahogany, into which had been carved a riot of figures: Saraswati, the goddess of knowledge; a lotus flower; a Buddhist swastika; a Nepali flag; and a Rotary wheel.

DURING HIS FIRST YEAR ON THE HILL, RAJEEV REALIZED THAT even the president has less power than Senator Patrick Leahy when it comes to Peace Corps funding. The president makes a budget proposal, which moves to the House, and then the Senate committee effectively has the final decision. And Leahy had become disillusioned with the Peace Corps bureaucracy during the George W. Bush administration. People told me that the agency's Washington office became so poorly managed that staffers often didn't know who Leahy was. "He would call the Peace Corps and say, 'This is Patrick Leahy,' and they would ask him to spell his name," Rajeev said. "And this is the guy in charge of funding them!"

Rajeev heard that Tim Rieser, Leahy's top aide for appropriations, was particularly critical. Rieser told me that the agency badly needs reform, especially in terms of directing more resources toward strategic countries. "The number of volunteers in Benin, a country so small that most people wouldn't know how to find it, is only about twenty-five less than China," Rieser said. "What we want to have a conversation about is: Does that make sense?" He emphasized that the appropriations committee works with a limited budget. "Every dollar for the Peace Corps comes out of something else," he said. "It comes out of programs for water, for food aid, for refugee resettlement." Rajeev told me that Rieser was right about the Peace Corps office's needing reform, but he believed that the new director, Aaron Williams, who was appointed in the summer of 2009, was committed to making changes. Williams arrived after a successful career as a USAID administrator, and he told me that one of his top priorities was evaluating where volunteers should be sent. In any case, Rajeev thought that Leahy's perspective would change if he spent more time in the field. "You can't judge the Peace Corps by what you see in Washington," Rajeev said. He wanted funding to be taken away from military programs, not aid. "Can you imagine how much bureaucratic waste there is when you give all this money to Pakistan and other foreign militaries?" he said.

But Rajeev couldn't figure out how to get to Leahy. He knew that the senator likes the Grateful Dead, and he's such a Batman fan that he had a cameo in *The Dark Knight*. "I talked to a longtime aide to Senator Kennedy, and I asked what I should do to convince Leahy," Rajeev told me. "She said, 'Dress up in a Batman suit and stand outside the Senate.' She was serious. I thought about it." Rajeev was able to persuade Jimmy Carter to call Leahy, but he couldn't get a statement from the Grateful Dead. He called everybody on a list of two hundred former volunteers who live in Vermont, asking them to telephone the senator. One of them happened to be the CEO of the hospice that provided care for Leahy's father when he was dying, and she agreed to make a call.

By this time, Rajeev's relationship with the National Peace Corps Association had begun to deteriorate. The organization wouldn't comment, but I heard from others that it became uncomfortable with Rajeev's tactics, especially after he and others began publishing editorials that targeted Leahy. Washington people responded in vastly different ways to this unorthodox approach. Republicans could be surprisingly supportive; they seemed to like Rajeev's individualism and outsider status. Elected officials and other important figures often admired his gall. Negative responses seemed most likely to come from younger people, especially congressional aides. "He makes things personal," one of them told me, and others complained that he hassled them and didn't follow the rules. Rajeev told me that each congressional office functions like a miniature village, with complex relations between the various aides and the elected official. In these villages, generational stereotypes are reversed: often, the younger people are most conservatively committed to a system, because they run day-to-day affairs. Marc Hanson, a former staffer for Representative Sam Farr, told me that officials tend to relish the personal and spontaneous side of politics, whereas staff members are focused on logistical issues. "They try to manage and choreograph that stuff," he said. "And Rajeev interrupted that process and forced members to hear about the Peace Corps on a day when it wasn't on the schedule."

When people discussed the Peace Corps campaign, they often said, in ominous tones, "You know about the ice-cream social, right?" I heard about it again and again, until the words "ice-cream social" began to sound like "Bay of Pigs." Each summer, there's a Washington fund-raiser for Patrick Leahy, featuring Ben & Jerry's ice cream and other Vermont products. At the time of the 2009 event, the Peace Corps's budget had reached a critical moment. Obama had asked for $373 million, a modest increase over the current $340 million, but the House, responding to the campaign, requested $450 million. Rajeev and Laurence Leamer, a former volunteer in Nepal and an author, attended the ice-cream social. In front of a large group of people, Leamer introduced himself to Leahy.

"Senator, I'm afraid I've got something to say that's not going to make you happy," he said, and then he read from a statement: "You should listen to your true progressive heart and not your negative aides." Leahy responded angrily, saying that he relied on his own judgment and was tired of people hassling him. But Leamer refused to back down. He said, "Senator, that's what democracy is all about."

An aide finally stepped between the men and ended the exchange. Leamer sent a written apology, but people involved in the campaign sensed that a line had been crossed. As Rieser told me recently, "It's not a smart way to start a conversation with the person in charge of funding." Not long after the social, the budget came in at $400 million—the largest single-year increase ever, but far less than the $450 million that had been approved by the House. A member of the House told me that Leahy complained to him about the rudeness of former volunteers, saying, "That'll cost the Peace Corps fifty million dollars." Leahy denies this.

Most people viewed the campaign as a resounding success, but Rajeev had mixed feelings. After the ice-cream social, his relations with the NPCA became so bad that he left the organization. For the next year, he worked independently, supported by grants arranged by a group of advisers led by Donald Ross, the former Nader activist. He used a less aggressive approach, and the campaign resulted in the Senate's proposing a twenty-million-dollar increase for the 2011 budget. But Rajeev believed that critical energy had been lost; he had hoped to have a bigger effect on the Peace Corps itself, whose reforms have been modest to this point.

Rajeev told me that he had taken the personal approach too far. Village politics can be nondemocratic: there's a point at which a powerful individual hears so many unified voices that his instinct is to ignore them. Leahy had been called by everybody from his father's hospice administrator to a former president, and he was confronted at the Ben & Jerry's ice-cream social. As if that weren't enough, he also received phone calls from both Ben and Jerry. Ben Cohen told me that, after being contacted by Rajeev, he called Leahy and suggested putting more money into the Peace Corps instead of the Pentagon. I asked Cohen how the senator had responded. "His response was 'Enough already! Everybody and his brother has been calling me about this!' "

IN SEPTEMBER, I ACCOMPANIED RAJEEV ON A TRIP TO NEPAL. On the way to Namje, we stopped in the nearby town of Bhedetar, where a man named Mani Tamang approached us. Mani had piercing eyes and a dark gaunt face, and he introduced himself by saying that he had been the guerilla commander of local Maoist forces during the years of unrest. In 2008, the monarchy was abolished, and the UN has been monitoring the country's peace process. The Maoists are now a legitimate political party, but they have not been aboveground for long and Mani seemed nervous. He wore a dirty T-shirt that said "Casual Style."

"Are you upset with me for some reason?" Rajeev asked.

"I've said thank you so many times that 'thank you' cannot describe how we feel," Mani said. "All over the district we have no buildings like the school you built."

"Do you have any criticisms?"

Mani believed that the Maoists had been excluded from development work. "When you were building the first water project I really wanted to meet you," he said. "But the people in

Namje wouldn't let me." He asked Rajeev if he would organize a water project in a nearby community, but Rajeev explained that he wasn't doing that work anymore. "I can raise the idea with the district irrigation officer," he said.

Rajeev didn't want to coordinate more infrastructure projects. Other communities were building their own water systems, which he thought was better. "If you do something well, then other people copy it," he said. "You don't need to do it on a huge scale." He had stopped building schools, too. He saw his role as ever-changing, and he was highly critical of himself. Some projects had failed entirely, like the woolen hats, and the women's co-op had yet to figure out a productive endeavor. Rajeev believed that the school buildings were too utilitarian; at one dedication, he gave a speech in which he said that the building "looks like a jail." He told the villagers that they should paint it in brighter colors. Even the water project had some negative effects. Water allowed people to build in cement, and villagers had embarked on a chaotic construction phase. Outside investors had moved in, realizing that Namje could become a resort town for people to escape the heat of the Terai Plain. They had bought so much land that prices had risen tenfold in the span of a year, and locals worried about losing cohesion.

Among charities, replicability is a key goal, but Rajeev didn't see the point of a development worker's repeating the same thing in many areas without ever sticking around to see the long-term results. He believed that construction projects tend to be overvalued, when in fact it's more important to spend time in one community as it moves forward. He said the access to water in Namje wasn't nearly as important as the way that villagers had learned to work together. "People often ask me, 'How many schools has the Peace Corps built? How many hospitals?'" he said. "The Peace Corps has been very good about not playing that game. But it's part of why the organization is still small, and why people don't know as much about it." In the Peace Corps, whose twenty-seven-month commitment is longer than that for most service organizations, volunteers often become ambivalent about traditional development work. They're more likely to see the complexities of change, and less likely to cheerlead for big projects and sweeping plans. The novelist Paul Theroux, who served in Malawi in the 1960s, has written critically about the NGO presence in Africa, and he told me that great ambitions tend to be destructive. He was more positive about small projects he had seen in Costa Rica which were organized at the village level. "We need to inspire people, not intimidate them," he said. "There's something about all aid that is somehow subversive."

When I spoke with former volunteers, they invariably said, "I got so much more out of the experience than I gave." It was also common to hear that the Peace Corps benefits the United States more than the rest of the world. I didn't really believe these sentiments—they seemed to be a way of expressing humility and respect. I had always liked the slightly subversive element of the Peace Corps, because it tended to be quiet and personal. But it seemed that the failure of the Peace Corps is that former volunteers rarely play the same outsider role back home, at least politically. Somebody like Rajeev could go from Namje to Congress, where he saw the place through new eyes, which made his presence disruptive. He was unusual, though. The United States is very good at shaking up the rest of the world, but it's all but impervious to anything that moves in the other direction.

ON OUR LAST DAY IN NAMJE, WE ATTENDED THE DEDICATION of a new building for agricultural training. A year earlier, Rajeev had raised money to buy five acres of prime mountaintop land, and the village hoped to become a center where Nepalis could study organic farming. Namje had enough water to handle the growth, and villagers were becoming ambitious. They had recently organized their own fund-raiser in the city of Dharan, coming away with the staggering sum of $150,000. But plans were even bigger: they wanted $1 million from the Nepali government, so that they could expand the new training center into a real college.

Tanka Bhujel had organized everything for the dedication event. There were big tents and banners, and officials came from all over the district. They received elaborate ceremonial gifts of Buddha statues and Gurkha knives. Rajeev gave a speech about development, and the audience cheered every time he mentioned money.

The next day in Kathmandu, we went to the office of Rakam Chemjong, the minister for Peace and Reconstruction. Tanka Bhujel had prepared three letters requesting government support for the new college, and he asked Rajeev to deliver them. But Rajeev seemed distracted; he told me he had trouble processing the extravagance of yesterday's event, and he wasn't sure what to say to the minister.

We met in a high-ceilinged room, where the minister greeted Rajeev warmly. He was accompanied by a couple of aides, known as *peons*, one of the more satisfying loanwords in the Nepali language. There was also a Maoist official who was helping to draft Nepal's new constitution. Rajeev began his presentation, but instead of asking for support he said that it would be a mistake to build a full college in Namje. First, the villagers needed to focus on their plot of land. "They will listen if you say a few words to them," he said.

"I will tell them to be careful and go slowly," the minister said:

"There's a lot of danger in the community right now," Rajeev said. "When I was a Peace Corps volunteer half an acre of land was three hundred dollars. Now it's ten thousand. It's like the Wild West."

"I understand," the minister said. The Maoist's cell phone rang—it played "The Internationale." He cut it off before the part about "the damned of the earth."

"We need to do something different in Namje," Rajeev said. "Something unusual."

The minister promised to keep an eye on the project, and everybody shook hands. Rajeev and I went outside and caught a cab. He still had the three request letters, undelivered, and I asked if Tanka and the others would be upset.

"They know me," he said. "Tanka Sir is amused by things like this. It will be a story for him to tell: 'We sent Rajeev in there with all these letters and he turned it around on us!'" The cabbie honked; we swerved through traffic, and Rajeev laughed: "He's the one who says that politics is the dirty game."

# BEACH SUMMIT

MY TROUBLE WITH THE POLICE STARTED AT LIN BIAO'S FORMER mansion. It didn't feel like a risky spot—the house was empty, and officially nothing was happening in Beidaihe. But everybody knew that the government had come to town. Beidaihe is a beach resort on the Bo Hai Sea, about two hundred miles east of Beijing, and officials had always gone there for vacation. Sometimes they went for work—each summer, the top Communist leaders met in Beidaihe and held secret conferences that helped determine the country's direction. The domestic press never reported on these meetings until they were over, but I could see signs that important people were in town. Haitan Road, where a number of powerful cadres had summer villas, was closed to traffic. Police stood on virtually every street corner. Periodically, a group of black Mercedes sedans roared across town, escorted by cops with blaring sirens; the convoy swept through and left the streets quiet in its wake, like a sudden summer squall.

It was 2002, and the meetings had started earlier than usual, in late July instead of August. Some analysts believed that this was significant—because so little about Beidaihe was made public, there was a tendency to search for meaning in every obscure detail. And there was no question that the stakes were higher than usual that summer. Jiang Zemin, the nation's leader, was expected to give up at least one of the three spots he held: general secretary of the party, head of the armed forces, and president of the People's Republic. But there were reports that Jiang and other members of the Old Guard were resisting retirement, and experts said that the Beidaihe meetings would be the first battleground of the political transition. Communist China had never had an orderly succession—for half a century, every transfer of power had involved coups or power struggles.

Lin Biao's mansion sat at the top of Lianfeng Mountain Park, like a monument to the hazards of power in China. During the early years of the Cultural Revolution, in the 1960s, Marshal Lin Biao was Mao Zedong's anointed successor, and his summer retreat reflected his status; it was famous for having a heated indoor swimming pool. But in September of 1971, after allegedly organizing a failed coup, Lin tried to flee China. In desperation, he drove out of Beidaihe—a soldier fired at the limousine as it left—and boarded a military plane in the nearby town of Shanhaiguan. Supposedly Lin Biao was en route to the Soviet Union when the plane crashed in Mongolia, although many details of this story remain unclear. In the years since his death, Lin has been portrayed as China's worst traitor, and also as a very strange man. According to Mao's personal physician, Li Zhishui, Lin Biao was terrified of wind, light, and water. He refused to drink anything; his wife dipped his steamed rolls in water to prevent him from getting dehydrated. That detail made me wonder about the heated swimming pool. The doctor also reported that Lin could move his bowels only if he placed a pan atop his bed, squatted above that, and covered himself with a tentlike blanket. Everybody I met in Beidaihe told me that I should take a look at Lin Biao's former home.

The sprawling gray brick mansion was in bad shape. Lianfeng was a public park, but the house was surrounded by an eight-foot-high wall, and visitors weren't allowed inside. Red paint peeled from around the windows and the sun had bleached the roof tiles to the color of sand. Two lightning rods poked above the surrounding pines. While I was there, a Chinese tour group passed by. Their guide was talking about the heated pool, too.

On the way back down the hill, I stopped to rest in a shaded spot, where I jotted a few sentences in my notebook. A young man in his early twenties walked past and asked what I was writing.

"Just my diary," I said.

"Can I see it?" he said, more urgently, and now I recognized something familiar about his manner. I put the notebook in my pocket. "It's nothing," I said. "I was just resting for a moment. I'm leaving now."

I made it fifty feet before he pulled out a badge—plainclothes cop.

"Let me see your notebook," he said.

"This is a public park," I said. "I haven't done anything wrong. My visa's fine."

I showed him my passport and then walked toward the exit. I was angry at myself—I knew better than to take notes in a sensitive place. But I had never had police trouble in a public park, and seeing the tour group had made me drop my guard. Now the cop followed close, talking on his cell phone. I kept my head down and walked out the gate; I heard the sound of people running to my left, and then three uniformed soldiers appeared and blocked my way. We stood facing each other in the parking lot. The soldiers were skinny men in their early twenties, and they looked nervous about the prospect of dealing with a foreigner. Three more jogged over from across the parking lot and cut off my retreat. Another plainclothes cop joined the first.

"We need to see your notebook," he said again.

"I'm not going to give it to you," I said. "There's no reason you need to see it." The book contained nothing sensitive, but I'd written down the contact information for some Chinese I'd met earlier in the day. They weren't involved in anything that could get them in trouble, but I knew the police might track them down and frighten them with a lot of questions.

Within ten minutes, three officials from the local foreign affairs department had arrived in a black Chinese-made Audi with tinted windows. The leader flashed his ID. I handed him my passport, and he recognized the "J" on the visa, which meant that I was a registered journalist. I told the officer that this was a public park and I hadn't broken any laws.

"We're not arresting you," he said. "We're just asking you to wait a minute."

"If you want to search me, then you'll have to arrest me," I said. "And I'll call the U.S. embassy to report the arrest of an American citizen."

I took out my cell phone. It was a bluff; I didn't know anybody at the embassy, and in any case my notebook wouldn't be a high priority for the State Department. The officials stood thirty feet away, talking in low voices. A couple of them made phone calls. After a few minutes, the head foreign affairs officer walked over.

"It's your choice," he said. "They just want to look at your notebook. It's no big deal. You can show it to them or you can refuse."

I told him I was keeping the notebook.

"Well," he said. "Then you can leave now."

I walked away without looking back. I figured I'd see them again.

BEIDAIHE SEEMED LIKE A PEACEFUL TOWN. THE SEA WAS GRAY and calm, shimmering in the flat northern light, and the streets were lined with willows and plum trees. Westward along the coast, the beaches were bureaucratically divided: one strip was reserved for vacationing soldiers of the People's Liberation Army; another section belonged to the State Council; a third was the private beach of the Diplomatic Services Bureau. After that came a wedge of sand that was open to anybody who paid a dollar's admission, and then there was the bustle of the free public beach: photograph vendors, drink sellers, and people who rented out shaded bamboo beach chairs. Women in skirted swimsuits waded tentatively into the water; men tucked cigarette packages into the waistbands of their Speedo-style trunks. The public beach stretched for half a mile and then ended abruptly at a roped-off border patrolled by uniformed guards.

Beyond the rope and the guards, the sand was reserved for the government leaders. In the mornings, I would wander over to see if anything was happening, but that beach always looked empty. The first indication of the leadership's presence came during the last week of July, when newspapers reported that Li Peng, the legislative chief, had been in Beidaihe when he received the speaker of the Maltese House of Representatives. But the reports said nothing about whether Li Peng was in town for the annual summer meetings. They noted only that Chinese-Maltese bilateral ties were stronger than ever.

Over a century ago, foreigners had started the tradition of using Beidaihe as a summer resort, building the first beach houses in town. After the Communists came to power in 1949, they converted Beidaihe into a party-run vacation spot. From the beginning it was intended to serve both government officials and the average employees of state work units. This followed the core theory of the revolution, which hoped to minimize the distance between a cadre and a worker. For decades, model employees were rewarded with free trips to Beidaihe—a week in the sun for a lathe operator or a ditch digger. Nowadays, most of the town's two million annual visitors were independent vacationers, but the coast was still home to government-run spas whose names had the ring of an earlier age: the Tianjin Teachers' Sanatorium, the Railways Cadre Resort.

I had checked in for a week at the Sanatorium for Chinese

Coal Miners. The resort opened in 1950 to provide both treat-ment and vacation accommodation for miners, who were heroes of the proletariat. Half a century later, they were still coming—every morning, I watched packs of men wander out from the sanatorium and stare wide-eyed at the sea; most of them came from mining towns deep in the interior. There were also private guests from other occupations who were paying for treatment at the sanatorium hospital, which specialized in necrosis of the hip joint. These patients included a tax bureau official from the northern province of Heilongjiang, a couple of oil-field workers from Daqing, and a woman journalist from the Shanghai bureau of the *People's Daily*. A retired postal worker told me that he had fought in Lin Biao's Thirty-ninth Army during the Korean War. He said that in 1950, not far from the Yalu River, they had sustained heavy casualties but held firm against MacArthur's troops. He told me the same thing I'd heard from other Chinese veterans of that war: "Americans can't eat bitter. They aren't as tough as Chinese."

I liked listening to the patients chat in the late afternoon. They sat in the shade in front of the clinic, enjoying the ocean breeze, their crutches propped nearby. Conversations drifted idly, as rhythmic as the hum of the cicadas in the plum trees. Occasionally they discussed politics; once, I asked about the meetings across town. "When it's done, they'll have reports in the paper," the tax bureau official said. "But until then it's not our business."

On my first day at the sanatorium, I met a seventy-two-year-old ethnic Russian named Sergei, who had suffered a stroke and was receiving therapy. His left side was partly paralyzed—his hand was strapped to a paddle to prevent the fist from clench-ing. He was in a wheelchair. He told me that his family had fled Siberia in 1938, and since then he had never left China. He spoke perfect Chinese and he had been a member of the Communist Party for fifty-two years. When I asked him why he had left Russia, he sighed. "That's a long story," he said, and then he slipped into silence.

A couple of days later, another patient asked Sergei the same question, but this time he was more forthcoming. He said that in the 1920s and 1930s the Japanese had sent secret agents to Siberia, and Stalin had ordered the expulsion of all Asians in the region. Sergei's parents were poor Russian farmers, but they had befriended some Chinese who were doing trade in Siberia. In 1938, during the actions against foreigners, the Chinese persuaded Sergei's family to move south with them, because conditions in Siberia had become so bad. "Stalin had some successes, but he also made some mistakes," Sergei explained.

"Like Mao Zedong," said the *People's Daily* reporter.

"Lenin was better than Stalin," somebody else said. "Lenin didn't make big mistakes."

"Wasn't Lenin a Jew?" a third person asked.

"Lenin was Russian. He wasn't a Jew."

"Why is it that Russian women are pretty when they're young, but then they get so fat?"

"I thought Lenin was a Jew."

"It's the way they eat."

"They eat soup before dinner instead of after."

"Russian women get fat because they don't care," Sergei said authoritatively. He was six feet tall and thin as a birch branch. His wife was Chinese; before they met, she had been in a dance troupe that entertained soldiers near the front lines during the Korean War. An explosion during a battle had left her partly deaf. "Russian women just don't worry about getting fat," Sergei said.

AFTER THE INCIDENT WITH THE PLAINCLOTHES COP, I HAD lunch at a noodle restaurant not far from Lin Biao's mansion. I took my time with the meal, hoping that the trouble would blow over, but after a while I noticed a young man watching me from across the street. When I walked outside, he stood up immediately and made a call on his cell phone. I thought about taking a cab far across town, to make it hard for anybody to follow me, but then I realized that it probably wasn't necessary. I had registered with the front desk when I arrived at the hotel, so they knew that I was a journalist, and all Chinese hotels were required to report foreign guests to the local police. It was almost certain that the officer I'd met at the park already knew where I was staying.

I waved down a cab and returned to the hotel. The Sanatorium for Chinese Coal Miners was a large, wooded complex with more than two dozen buildings, and I was staying in the old VIP quarters. They had opened it up to foreigners a few years earlier, and a suite cost about forty dollars a night, with breakfast included. But I never saw any other Western guests; the cavernous building was mostly empty. Out in front, there was a weed-filled lawn with old marble goats that had been stained gray by the weather and the sea breeze. As I approached, I noticed a man standing near one of the goats. He took out his cell phone when I came into view, and then he walked into the lobby ahead of me.

My room was on the second floor. On the way to the stairs, I turned and caught another glimpse of him: crew cut, dark polyester slacks. He talked quietly into his phone, staring at me.

I waited in my room. Whenever I had trouble in a Chinese city, it was only a matter of time before a foreign affairs officer came to kick me out. Technically, a foreign journalist was supposed to apply in advance before going anywhere, but in practice nobody obeyed those rules anymore, and usually I traveled undisturbed. On the rare occasions when I attracted police attention, though, the response was typically swift. There would be a knock at the door, and an official with a tight smile would tell me politely that he would be happy to have me visit again in the future, provided that I apply to his office. In the meantime I should return to Beijing.

My two-room suite had a musty smell, but the light was good and there was a balcony. Next to the telephone, a "Services Offered" manual listed damage fines for practically every object in the room. Ashtrays cost 12 cents each. A teacup was 61 cents, and a towel was $1.82. Ruined carpet went for $6.05 per square meter. If I smashed the mirror, I'd have to pay $12.11. The most expensive thing in the room was the toilet, which cost $60.53.

After thirty minutes of waiting for the knock on the door, I fell asleep. The confrontation over the notebook had left me exhausted, and I slept for more than an hour. I awoke disoriented, and then I remembered where I was: the $3.63 pillow, the $4.84 sheet. For some reason there wasn't a price for the bed. I went downstairs and the crew-cut man was still in the lobby, cooling himself with a bamboo fan. He saw me and froze, the fan directly in front of his face.

THERE WASN'T MUCH TALK IN BEIDAIHE ABOUT THE SUCCESSION. A couple of times, I had a conversation with an educated person who mentioned Hu Jintao, the current vice president. Most experts believed that Hu was the most likely candidate to succeed Jiang Zemin if the older man actually retired. But it was rare to hear anybody in Beidaihe say much about it. The subject didn't make them nervous or wary; they simply felt that it wasn't their business, and they didn't expect that a change in the leadership would affect their lives. When they did talk about it, they tended to do so metaphorically, as if this made the issue more relevant. Sergei compared the transfer of national power to what happens in a family. "Say the father gets old and puts the eldest son in charge," he said. "It's not right for the younger sons to resist him. Unless he makes some major mistake, he should remain in charge." Sergei had been a low-level cadre—he'd served for three years in the People's Party Congress in the western city of Urumqi—and he said that China should learn from the mistakes that Russia had made during perestroika. "Gorbachev was too much in a hurry," he said. "They should have reformed the economy first and gone slower with the politics."

Other than Sergei, the most outgoing patients in the sanatorium were Yao Yongjun and Zou Yunjun. They came from the oil city of Daqing, in northeastern China, and they both had bad hips. In the 1960s, Daqing was a national model for Chinese industry, but the transition to the private economy had been rough in that part of the country. Recently, there had been big layoffs in the state-owned oil company, and some of the workers had demonstrated. Yao and Zou said that their work units hadn't been affected yet. Each of them still earned roughly two hundred dollars a month, a good wage, and they told me that they liked their jobs. Yao was a fat thirty-two-year-old with a quick laugh; he had nearly finished his two-month treatment, which involved a combination of traditional Chinese medicine, injections, and ultrasound. He told me that necrosis of the hip can be caused by three factors: hormone imbalances, excessive drinking, and trauma. He believed that the first two had brought about his hip problem.

Zou Yunjun said that in his case it was a matter of all three. He was thirty-nine years old and he did manual labor on the oil fields. Zou had a crew cut and a scar that started at his hairline, ran straight down his forehead, and ended in a point between his eyebrows, like an exclamation mark. His given name was a combination of the words "cloud" and "army," and he said that it meant "Army of the Heavens." His father had given him the name because of his military experience—he was another Korean War veteran who had fought near the Yalu River. Zou was well built and he often went shirtless around the sanatorium. A gold chain hung around his neck. He wore a fake Rolex. He flirted with the nurses, and they seemed to like it.

One afternoon, I was sitting in the shade with Zou, and a nurse told him to come for treatment at eight o'clock the following morning.

"That's too early," Zou said, and he grinned. "Anyway, I like to use the toilet at eight o'clock."

The nurse giggled and covered her mouth with her hand. "OK," she said. "Nine o'clock."

"Ten," said Zou.

The nurse scurried off, laughing uncontrollably. Later that day, I asked Zou about politics, and he said that the Chinese leaders he most admired were Mao Zedong; Tang Taizong, the second emperor of the Tang dynasty; and Genghis Khan.

"But he was Mongolian," I said.

"Mongols are one of the fifty-six nationalities of China," Zou said. He told me that he admired Genghis Khan because he had expanded the Chinese empire all the way to Moscow. I knew that very few Mongolians would agree that their national hero had been Chinese, but this was a common belief in China. I asked Zou what he thought about Jiang Zemin.

"He's fine," he said, shrugging. "But it's hard for him to compare with all these famous leaders from history."

Zou and the other patients didn't seem to notice the plain-clothes cops. There was always at least one hanging around the lobby of the VIP building, and whenever I walked on the grounds of the sanatorium, somebody moved in the same direction, keeping me in sight. Outside the front gate, as many as four men waited; if I left the complex, they activated their cell phones. I had never experienced anything like this in China—usually, if the authorities wanted you to leave, they simply asked. But here they seemed content to let me wander freely as long as I remained in sight. One morning, I finally approached two of the cops at the gate.

They tried to appear nonchalant. One was the crew-cut man I had seen in the lobby on the first day—heavyset, in his forties, wearing a dirty tan T-shirt. His partner was better dressed: a fake Izod shirt and black leather loafers that featured the *Playboy* logo. He had bad skin. I asked him where he was from.

"Changchun," Playboy said, naming a city in the northeast. He said that he was in Beidaihe for therapy, and I asked him what was the matter.

"Nothing's wrong," he said quickly. "I'm just resting. It's vacation."

"I thought you said you were here for therapy."

"That's not what I meant. I'm here for vacation." He commented awkwardly on Beidaihe's good weather and nice beaches, and then he said he had to leave. The two of them wandered off, Playboy and Crew Cut; they cast furtive glances over their shoulders. It was one of the few conversations I'd had with a Chinese in which I wasn't asked a single question.

In Beidaihe, it was tempting to see people as politically naive. Each summer, the nation's leaders came to town, and some of the most important decisions in the history of the People's Republic had been negotiated here. Beidaihe had been home to the final drama of Lin Biao's career, when the highest leadership of the Communist Party had been at risk of a coup. But none of this seemed to matter to people on the street. Lin Biao's house was fenced off, and apart from the heated pool, the mansion didn't inspire conversation. In the same way, people showed no curiosity about Jiang Zemin and Hu Jintao, or about the upcoming transition.

But it was also true that the average Chinese citizen had very little information about these men. Hu Jintao was a cipher—in the Communist Party, it was a tradition for up-and-coming leaders to maintain a low profile. Hu was fifty-nine years old, and he had spent much of his career in the Chinese hinterlands. He never granted interviews. His most significant political experience had been in Tibet, where he had been appointed to Communist Party secretary, the top position, in 1988. That had been a sensitive moment—the region had been suffering from ethnic unrest, and the previous leadership had just been purged.

One of Hu's first significant acts was to give the eulogy at the funeral of the Panchen Lama, the most important Tibetan holy man after the Dalai Lama. Some Tibetans believed that the Chinese government had arranged for the Panchen Lama to be murdered, and tensions were so high that nothing was going to be solved with a eulogy. Hu quoted Deng Xiaoping's praise of the Panchen Lama as a staunch patriot of China. Within a month, after dozens of Tibetans died in violent clashes with the police, Tibet was placed under martial law, which lasted for two years. During this time, Hu was neither particularly cruel nor particularly skillful. He simply followed Beijing's orders and rode out the storm.

In the same way, Jiang Zemin was best known for surviving difficult times. In 1989, when he was the party secretary of Shanghai, he was able to keep the city mostly peaceful during the Tiananmen Square protests and subsequent crackdown. As president of the nation, he weathered the Asian financial crisis. But Jiang had little charisma and he never captured the popular imagination; unlike previous leaders, he was neither a war hero nor a former peasant. He seemed to share much of his power with others in the Politburo. Jiang wore heavy, old-fashioned glasses, and sometimes he made awkward efforts to deliberately echo the words of his more revered predecessors, Mao and Deng. His contribution to the nation's ideology was a twenty-thousand-word speech about development that became known as "Three Represents." Millions of Chinese studied this document in their schools and work units, but very few of them could tell you anything meaningful about it. Much of the speech was an exercise in tautology: "All relations of production and superstructures, regardless of their nature, develop with the development of productive forces. . . . As for how things will develop specifically in the future, the answer to this question should come from practice in the future." The language of the speech became clear only when it defined a negative: "We must resolutely resist the impact of Western political models such as the multiparty system or separation of powers among the executive, legislative, and judicial branches."

In the same way, it was easiest to define Jiang Zemin and Hu Jintao by who they were not. They were not Mao Zedong, and they were not Mikhail Gorbachev. China was no longer ruled by one man's whim, and it was not yet ruled by law. The People's Republic was the first Communist state to have outlived the cult of personality, and the blandness of men like Jiang and Hu was perhaps their most salient characteristic. Not long after the Beidaihe conference, it was announced that Jiang Zemin would resign from all three of his leadership positions—the first Chinese Communist leader to willingly step down. During Hu Jintao's first year in power, he ended the tradition of meeting secretly in Beidaihe. The beach summits belonged to an earlier era, when individual leaders mattered much more, and when there always seemed to be the risk of a purge or a coup. Today's China had become much more systematized, and its totalitarianism had evolved into something else: a one-party state with a high degree of economic freedom. If the citizens seemed passive, it was because they had seen much worse. After all those years it was a relief to think about something other than politics.

THE DAY THAT YAO YONGJUN'S HIP TREATMENT WAS FINISHED, he telephoned and invited me to go out drinking to celebrate. I didn't know whether I should go—the oil workers seemed like simple men, and I was afraid that my trouble with the cops would somehow compromise them. I had been careful to limit my contact with the patients to the public parts of the sanatorium. After thinking about the invitation, I decided that it was best to explain my concerns to Yao and see how he responded.

But as I was walking through the sanatorium to meet him, Playboy appeared and took a seat near some of the patients. Crew Cut came from another direction and stood nearby. Nobody else seemed to notice them; the patients chatted as usual. In recent days, I had felt myself slipping into a strange private world: much of my attention was focused on the surveillance, and yet life around me continued as usual.

Yao Yongjun hobbled over and shook my hand, smiling broadly. Two of the other patients pulled up in a car and told us to hurry, because everybody else was waiting. Reluctantly, I got inside. As we left, I saw Crew Cut put his phone to his ear. I imagined this wasn't a hard assignment: Look for a foreigner and three guys on crutches.

We hired a private room on the second floor of a restaurant. There were seven of us, the crutches lined up neatly against the wall. It was warm, and one by one the men took off their shirts in preparation for the drinking. A waitress served warm beer and bowls of ice cubes. There were many benefits to the years I'd spent in China, and one of them was that I'd grown to appreciate the way that beer fizzes when you toss an ice cube into the glass. A man at the table ordered a bottle of Diwang wine, which came from vineyards nearby. The Diwang didn't fizz when he added the ice.

One old man who worked in a coal mine was smoking Big Double Nine cigarettes. "That was Chiang Kai-shek's favorite brand," Zou Yunjun said.

"Do you know what Mao smoked?" Yao asked me.

I guessed the brand that is named after the central government compound in Beijing. "Zhongnanhai?"

"Zhonghua," Yao corrected me. "China brand."

"Deng Xiaoping smoked Panda cigarettes," somebody else said.

"Zhou Enlai smoked Zhadan."

I asked what Jiang Zemin smoked.

"He doesn't smoke. The leaders watch their health nowadays."

Zou changed the subject. "What's the difference between Chinese women and foreign women?" he asked.

After a while, I went downstairs to see if Playboy or Crew Cut was on the job, but I didn't see either of them. Back at the table, Yao Yongjun asked if I liked Beidaihe. We had been drinking steadily for more than an hour, exchanging toasts around the table, and his face was flushed. I chose my words carefully.

"To be honest, I've had some problems here because I'm a journalist," I said. "It's not usually this way, but there have been a lot of police following me. Actually, there were two plainclothes officers in front of the hospital when we left."

"I already know about that," Yao said.

"You know about the police?"

He nodded. His eyes were so calm that it took me a moment to continue.

"I don't want to cause you any trouble," I stammered.

"It's no trouble," he said. "It doesn't worry us. Anyway, there's nothing you can do about it," he added. "Just ignore it."

He made a toast to his hometown, which was halfway to Siberia. The room went silent: no murmurs, no cicadas. The only sound was the ice clinking as we raised our glasses to the oil fields of Daqing.

# BOOMTOWN GIRL

EMILY COULDN'T TELL ME EXACTLY WHY SHE HAD LEFT HER hometown after graduating from the local teachers college. "There was something in the heart," she said. "My mother says that I just won't be satisfied with a happy life. She says that I'm determined to *chiku*—eat bitter." In any event, she wouldn't have been content with life as a local schoolteacher. "Teaching is a good job for a woman, and it's easy to find a husband, because men like to have teachers as their wives," she said. "It could have been a very comfortable life. But if it's too comfortable I think it's like death."

She was slipping away even when I first met her. That was in 1996, when I taught English at the teachers college in Fuling, a small town on the Yangtze River in Sichuan Province. My students were training to become middle-school teachers, and one day I asked them to respond to a hypothetical question: Would you rather have a long life with the normal ups and downs, or an extremely happy life that ends after another twenty years?

Nearly all my students took the first option. Most of them came from farms in the Sichuan countryside, and several pointed out that their families were so poor that they couldn't afford to die in two decades. Emily, though, chose the short life. At nineteen she was the youngest student in the class. She wrote:

It seems to me that I haven't been really happy for quite a long time. Sometimes I owe my being dispirited to the surroundings, especially the oppressive atmosphere in our college. But I find the other students can enjoy themselves while I am complaining, so I think the problem is in myself.

Everything she wrote that year marked her as different. She contradicted her classmates; she skirted the Communist Party line; she had her own opinions. She wrote about her father, a math professor who had spent the Cultural Revolution in political exile, working in a coal mine; and she wrote about her older sister, who had gone to the city of Shenzhen, seven hundred miles away, to look for work. When I asked my students to compose a business letter to an American organization, Emily chose the Country Music Association, in Nashville. She told me that she was curious to learn what country music was like. Another time, she asked if I had any black friends, because she had never seen a black person, except on television. When my literature class performed *A Midsummer Night's Dream*, she played Titania. She was a good actress, although she had a tendency to play every role with a touch of a smile, as if she were watching herself from a great distance. She had a round face, high cheekbones, and full lips. Her eyes were thin and delicate, like those of a woman in a classical Chinese painting. Emily once remarked to me that her features had been considered beautiful when she was a young teenager, but nowadays the preference was for bigger eyes, because they looked more Western. She did not wear makeup. She dressed simply, and unlike many young Chinese women she did not dye her hair. She had chosen her English name in honor of Emily Brontë.

After graduation, most of my students accepted government-assigned teaching jobs in their hometowns. But Emily went south to Kunming, the capital of Yunnan Province, with her boyfriend to look for work. He was a square-faced young man with bristly hair, hard black eyes, and a quick temper, and he wanted to continue on to Shanghai. "I hadn't decided to break up with him yet," Emily told me later. "But I knew I didn't want to go to Shanghai." Instead, in November of 1997, she went to Shenzhen, and within a few months they had broken up for good. In Shenzhen, it took her less than a week to find a position as a secretary in a factory that produced costume jewelry for export. Her starting salary was 870 yuan a month, or $105 in U.S. dollars. Most of her former classmates were earning about $40 a month as teachers in Sichuan.

It was not unusual for former students to call and tell me about various milestones of independence. Often, these had to do with money and new possessions—a salary raise, a new apartment. Once, a student called to tell me that he had acquired a cell phone. He talked about the cell phone for a few minutes, and then he mentioned, in an offhand way, that he had also become engaged. Five months after starting the factory job, Emily called to report that she had received a raise, to $120 a month.

She laughed when I said that she now made as much money as I did. But she sounded a little funny, and I asked if something was wrong.

"The company has an agent in Hong Kong," she said slowly. "He often comes here to Shenzhen. He is an old man, and he likes me."

"What do you mean by that?"

"Because I am fat." She giggled nervously. I knew that she had gained a little weight, and in some ways it made her even prettier.

"Does he want you to be his girlfriend?"

"Perhaps." Her voice sounded small on the phone.

"Is he married?"

"He is divorced. He has young children in Taiwan. But he usually works in Hong Kong."

"How often does he come to Shenzhen?"

"Twice a month."

"Is it a big problem?"

"He always finds a way to be with me," she said. "He says he will help me find a job in Hong Kong if I want one. The salaries are much higher there, you know."

"That sounds like a very bad idea," I said carefully. "If you want another job, you should not ask him for help. That will only cause big problems in the future. You should try to avoid him."

"I do," she said. "And I tell my coworkers to always be with me if he is here."

"Well, if it becomes a big problem, you should leave the job."

"I know," she said. "Anyway, it is not such a good job, and if I have to leave I will."

IN THOSE DAYS, THE CITY OF SHENZHEN WAS SURROUNDED BY a sixty-seven-mile-long chain-link fence. It was about ten feet high, and some sections were topped by barbed wire. If you approached the city from the north, you entered one of the fence's checkpoints and followed a modern highway through low green hills. The new buildings grew taller as you approached the downtown area. At the intersection of Shennan and Hongling roads, there was a massive billboard that represented, at least in the spiritual sense, the heart of the city. The billboard featured an enormous image of Deng Xiaoping against a backdrop of the Shenzhen skyline, with the phrase PERSIST IN FOLLOWING THE COMMUNIST PARTY'S BASIC LINE FOR ONE HUNDRED YEARS WITHOUT CHANGE. Locals and visitors often posed for photographs in front of the sign. In February of 1997, when Deng died, thousands of Shenzhen residents spontaneously gathered at the billboard to make offerings of flowers, written verses, and other memorials. They sang "Spring Story," which was the official Shenzhen song:

*In the spring of 1979*
*An old man drew a circle on the southern coast of China*
*And city after city rose up like fairy tales*
*And mountains and mountains of gold*
  *gathered like a miracle . . . .*

Other Chinese cities have history, but Shenzhen's origins have the flavor of myth—the miraculous birth, the benevolent god. In 1978, two years after the death of Mao Zedong, Deng Xiaoping marked his rise to power by initiating what became known as Reform and Opening—capitalist-style innovations that ended almost three decades of Communist economics. Deng avoided trying out the more radical changes in major cities like Beijing and Shanghai, where mistakes would be politically disastrous. Instead, he and his advisers experimented in less developed areas, in what came to be called Special Economic Zones. Through a system of tax breaks and investment privileges, the government hoped to encourage foreign firms to set up shop in these zones. In 1980, they conferred this honor on Shenzhen, a sleepy southern border town whose economy had depended mostly on fishing and farming. Shenzhen became a "reform laboratory," and one of its nicknames was Window to the Outside World. Before long, major international corporations established plants in the city, including the American firms of IBM, Compaq, PepsiCo, and DuPont.

In 1990, the government established the Shenzhen Stock Exchange—the first big-board market in China. (The second one opened later that year, in Shanghai.) For more than twenty years, Shenzhen had an average annual growth rate of more than 30 percent, and its residents came to enjoy one of the highest standards of living in any Chinese city. The place was unusually green—parks were sprinkled throughout the downtown area, and streets were lined with banyan trees, palms, and well-groomed strips of grass. There were few bicycles downtown;

most people could afford to take buses or cabs. Traffic moved smoothly. The city center was intersected by Shennan Road, nine lanes of cars bordered by Shenzhen's best-known buildings: the Stock Exchange, a block of glistening blue-green glass; the Land King Tower, a narrow, twin-spired building of sixty-nine stories; and its adjoining apartment complex, which, with a seven-story-high opening in its facade, was the city's most architecturally innovative structure.

During its rise, Shenzhen became home to a social experiment that was even more impressive than its economic adventure. The city's population exploded from three hundred thousand in 1980 to more than four million in 2001. By then, the average Shenzhen resident was less than twenty-nine years old—a remarkable statistic in a country whose planned-birth policy was already resulting in aging urban populations. Because many factories relied on unskilled, low-wage labor, most of the newcomers were women. Although there were no reliable official statistics, locals liked to say that there were seven women to every man in Shenzhen. This was an exaggeration, but one that reflected the general trend. Sometimes people called the city a "woman's paradise," because it offered so many job opportunities to young women. But this phrase hardly described the underside of the boomtown. Shenzhen was also notorious as a place where workers in poorly regulated factories suffered injuries, sometimes losing limbs in labor-related accidents. The first phase of China's free-market development spurred an explosive growth in the sex industry, and nowhere was this phenomenon more obvious than in downtown Shenzhen, where it was impossible to walk at night without being propositioned by young prostitutes, who were known as "street angels."

Whenever China faced a moment of political uncertainty, Shenzhen felt like a city under siege. Locals knew that they had benefited from unusual government patronage, and they constantly worried that Shenzhen's special status might be revoked.

The 1997 offerings to Deng Xiaoping at the billboard reflected this fear of political change. In the mid-1980s, when a series of smuggling scandals broke out in the Special Economic Zones, conservative Communist officials targeted Shenzhen. They complained that the loosened restrictions on foreign investment were invitations to corruption and neocolonialism.

Eventually, such worries inspired the government to erect the fence around the city. It was a distinctly Chinese solution: just as the Great Wall had been built to keep foreigners at bay, so the Shenzhen fence was intended to keep capitalist reforms under control. Chinese citizens entering the city proper had to go through customs, where they showed a border pass and an ID that required approval from their home province. But the completion of the fence in 1984 had unintended consequences. Labor-intensive factories inside the Special Economic Zone began moving to the other side of the fence to take advantage of cheaper rents and less rigorous law enforcement. Eventually, Shenzhen became divided into two worlds, which were described by residents as *guannei* and *guanwai*—"within the customs" and "outside the customs." Satellite towns sprang up beyond the fence, most of them squalid and unplanned. In this sprawl of cheaply constructed factories and worker dormitories, wages were lower, and workers tended to rely heavily on overtime bonuses. Six-day workweeks were standard, as opposed to five-day weeks in Shenzhen proper. There were a lot more labor accidents and dormitory fires on the other side of the fence.

It was here that Emily found her first factory job. In the satellite city of Longhua, she worked as a secretary who handled inventory, kept track of orders, and did some English translation. Her factory exported costume jewelry—pieces of pewter and brass, and cheap plastic beads that were painted, lacquered, and packaged in ziplock bags to be sent to Hong Kong, to Southeast Asia, to San Francisco, to Chicago.

\* \* \*

HER STORIES DRIFTED UP FROM THE SOUTH. EVERY TWO OR three weeks, Emily telephoned or wrote, creating the city in my mind. Some of her tales ended abruptly, like the one about the businessman from Hong Kong who had pursued her. Other stories lasted longer, like the one about her older sister, who had first worked as a traveling saleswoman and then been recruited by a company that was running a pyramid scheme. She brought Emily to the recruitment meeting. "A lot of the salespeople had low cultural levels, but they had learned how to talk," Emily recalled. "I didn't think it was a good way to make money, but it was a good way to improve yourself and improve your confidence." Her sister had known that it was a scam—the government was cracking down on pyramid schemes, which had run rampant across southern China—and she said that she had gone to the meeting simply out of curiosity. Afterward, she had taken a job with a lonely-hearts hotline, talking on the telephone with other people who felt lost in Shenzhen. "Some people say there is no real love in Shenzhen," Emily said, when I asked why the hotline existed. "People are too busy with earning money to really live."

That was probably why a young man named Zhu Yunfeng took her by surprise. He had been trained as a mold maker, but he came to the jewelry factory as a purchasing agent, because he wanted a break from manual labor. At his previous job in Shenzhen, Zhu Yunfeng had miscalculated the weight of a metal part, which slipped when he and two other workers were trying to lift it. Zhu Yunfeng let go. The other two workers didn't, and they lost some of their fingers. The workers were promised compensation, and Zhu Yunfeng wasn't blamed for the accident, which was a relatively common occurrence in plants beyond Shenzhen's fence. Still, he decided to leave the job. Seeing the injured men around the factory made him feel uncomfortable.

At first, Emily didn't take much notice when Zhu Yunfeng arrived at her factory, in March of 1998. He was of average height, with stiff black hair, and his shoulders were broad from working with the molds. He wasn't handsome. He kept to himself, and none of the other women who worked at the plant thought that he was attractive. But after a while, Emily found that she was noticing him a lot more. She liked the way he walked—there was confidence in his gait.

Two months later, small gifts started appearing in the drawer of her desk. She received two dolls and a small figurine of a sheep. She didn't ask who had put them there.

One day in September, Zhu Yunfeng and Emily went out with some of their coworkers and found themselves walking alone in the local park. Emily didn't know how they had become separated from the group. Suddenly she felt afraid—things were happening too fast. She was twenty-two years old. He was twenty-six.

"I don't want to walk with you," she said.

"Who do you want to walk with?" Zhu Yunfeng asked.

"I don't want to walk with anybody!"

They turned around and went back to the factory. Months later he would tell her that that was the moment when he knew there was a chance of success. He could see that she still hadn't made up her mind to reject him.

The jewelry factory had fifty employees. The owner was Taiwanese, like many of the bosses who ran plants outside the customs, and he told the workers openly that he hated mainland China and was there only because of the cheap labor. The workers didn't like the Taiwanese owner very much. Some of them made as little as twelve cents an hour, which meant that they had to work overtime to earn a decent income. Whenever they talked about the boss, they used the same words that were commonly applied to Taiwanese owners in Shenzhen: "stingy" and "lecherous." But the jewelry-factory boss wasn't as bad as many of the others, and conditions at the plant were better than those at most factories beyond the fence. The workers had Sundays off, and during the week they were allowed to leave the factory after work hours, although everybody had to be back in the dormitory by eleven or midnight, depending on the boss's whim.

The dormitory where Emily lived was on the top two floors of a six-story building. There were six workers to a dorm room. It was a "three-in-one" factory—production, warehousing, and living quarters were combined into one structure. This arrangement was illegal in China, and the workers knew it, just as they knew that some of the production material stored on the first floor was extremely flammable. They also knew that the building had bad wiring, because an electrician had come to make a repair and commented to Emily, in an offhand way, that the place could go up in flames. Afterward, she mapped out an escape route for herself. If a fire broke out at night, she would run to the dormitory's sixth-floor balcony and jump across to the roof of the building next door. That was the extent of her plan—she had no interest in complaining to the government about the violations, and neither did the other workers. All of them were far from home, and they knew that such conditions were common in the plants outside the fence.

One Saturday night in October, Zhu Yunfeng took Emily's hand as they prepared to cross the road. She felt her heart leap up in her chest. Zhu Yunfeng held on tightly. They stepped into the street.

"I'm too nervous," Emily said, once they had reached the other side. "I don't want it to be like this."

"What's wrong?" Zhu Yunfeng said. "Haven't you ever done this before?"

"I have," she said. "But I'm still scared."

"It's going to be like this in the future," Zhu Yunfeng said. "You should get used to it."

\*    \*    \*

EMILY LAUGHED WHEN SHE TOLD ME THAT STORY, DURING MY first visit to Shenzhen. She had a certain gesture that she often made while laughing—she put her hand over her mouth and closed her eyes. Many Chinese women laughed like that, but for some reason it seemed more natural when Emily did it.

She looked much the same as I remembered from her student days. On the morning that I arrived, she wore a simple blue dress, and the two of us wandered around downtown Shenzhen. We visited the Stock Exchange and the Diwang, the tallest building in town, and I took Emily's photo in front of the Deng Xiaoping billboard. At the end of the day, we caught a local bus north: past the downtown skyscrapers and the apartment blocks, watching the neighborhoods grow seedier with every mile we traveled from the center. We cut through the green hills just before the border, and then we arrived at the long, low line of the chain-link fence. There was a billboard at the checkpoint: MISSION HILLS GOLF CLUB: THE FIRST 72-HOLE GOLF CLUB IN CHINA.

Beyond the fence was a rough cluster of unfinished concrete buildings. Piles of dirt stood beside enormous foundation holes; bulldozers and dump trucks were parked near makeshift shacks where construction workers lived. We kept going north, the factory towns appearing one after the other: dormitories surrounded by fences, smokestacks sprouting in dirty clumps. Signs above factory gateways identified the goods produced within: shoes, furniture, toys, computer parts. After twenty miles, we arrived in Longhua, where a group of Taiwanese bosses had set up half a dozen jewelry plants. The buildings were arranged in a tight little group, as if something about this sprawling city had made them huddle close. A space of only a few feet separated Emily's factory and the neighboring one—this was why she had figured she could jump across to the roof if there was ever a fire.

That night we had dinner at an outdoor restaurant on the

town's main street. It was a pleasant evening; I had always preferred nighttime in China, when the flaws of the brand-new cities
tended to fade into darkness. This was particularly true in the
Shenzhen satellite towns, where most people spent their days on
assembly line shifts. Daytime streets tended to be empty—often
the place seemed almost abandoned. But the mood changed when
workers poured out of factory gates in late afternoon. Some of
them wore their uniforms, but most made an effort to dress up in
new clothes. They gathered in outdoor restaurants and pool halls,
and they moved in same-sex groups: packs of boys talking loudly,
groups of girls laughing together. It was rare to see a family or
even a child. Elderly people were essentially nonexistent. This
was the freedom of Shenzhen—there were no traditions here, and
no sense of the past; everybody was far from home.

Emily sent part of her salary back to her parents, to help
pay for her brother's school tuition, and the responsibility had
given her a new air of maturity. At twenty-three, she was the
oldest worker in her factory office, which was staffed entirely
by women. During dinner, she regaled me with stories about the
Taiwanese factory owners. They fascinated her—she couldn't
believe the looseness of their lives, financially and morally, and
she saw them as symbolic of the world beyond the mainland. She
told me about one of her boss's colleagues, a Chinese-American
who had recently arrived from San Francisco on business, had
faxed his wife a love letter from Emily's office, and then had
gone out and hired a prostitute. Emily's boss was always leering
at the young women in the factory; another Taiwanese businessman in a nearby plant had become so distracted by two Sichuanese mistresses that his company went bankrupt. In Emily's
opinion, they were all the same. "All of them have failed somewhere else," she scoffed, explaining that her boss's old company
in Taiwan had gone bankrupt years ago. When I asked if Shenzhen had fewer political restrictions than Fuling, she said it did,
but pointed out that the labor practices were just as restrictive.

"Here it's not the government but the bosses who control everything," she said. "Maybe it amounts to the same thing."

She told me about a notorious Taiwanese-owned purse factory in a nearby city. The plant kept the gates locked throughout the six-day workweek—except on Sundays, the workers couldn't leave the factory complex.

"That can't be legal," I said.

"Many of the factories do that," she said, shrugging. "All of them have good connections with the government."

She explained that one of her friends had worked at the purse factory, where the Taiwanese boss often ordered everyone to work until midnight, yelling when they got tired. One worker had complained and been fired; when he tried to claim his last paycheck, the boss had him beaten up. Emily decided that she had to do something about it, so she wrote the boss a letter that said, "This day next year will be your memorial day."

"And I drew a picture of a—" She was speaking English, and she couldn't think of the word. She pushed aside her plate and sketched an outline on the table—a simple head, a narrow body.

"A skeleton?" I asked.

"Yes," she said. "A skeleton. But I didn't write my name. I wrote, 'An unhappy worker.'"

I didn't know how to respond—back in Fuling, my writing class hadn't covered death threats. Finally I said, "Did the letter work?"

"I think it helped," she said. "Workers at the factory said that the boss was very worried about it. Afterward, he was a little better."

"Why didn't you complain to the police?"

"It doesn't do any good," she said. "All of them have connections. In Shenzhen, you have to take care of everything yourself."

When we finished the meal, Emily looked at me and said, "Do you want to see something interesting?"

She led me to a small street near the middle of town. Below the road, a creek flowed sluggishly in the shadows. Dozens of people stood along the curb, smoking cigarettes. The street was unlit, and all of the people were men. I asked what was going on.

"They're looking for prostitutes," Emily said. We watched as a woman appeared—walking slowly, glancing around, until a man came up and spoke to her. They talked for a few seconds, and then the man slipped back into the shadows. The woman kept walking. Emily said, "Do you want to see what happens if I leave you here alone?"

"No," I said. "We can leave now."

I spent the night in Zhu Yunfeng's one-room apartment. He had recently left Emily's factory to start a new job, which allowed him to live in private housing. His neighborhood was covered with boldfaced flyers advertising venereal disease clinics; we followed a string of them up the stairwell to Zhu Yunfeng's apartment on the fourth floor. The building was only half constructed—the walls unpainted, the plaster chipping away, the plumbing unfinished. They didn't have hot water yet and probably never would. Much of the development beyond the Shenzhen fence seemed to be like this—abandoned before it was completed. There was so much work to do, so many new factories and apartments to build, and contractors moved on once the bare essentials were in place. It occurred to me that if anything in this region was actually finished, it was immediately sent away for export.

Zhu Yunfeng's apartment was furnished with two simple wooden beds covered with rattan mats. There was nothing on the walls, and apart from a thermos and a few books, he didn't have many possessions. His current job involved making molds for the production of household appliances.

I knew that something about Zhu Yunfeng made Emily feel secure. Once, she had bluntly told me that he wasn't handsome, and this was true—acne had badly scarred his face. But his plainness was attractive to her. She had a theory that handsome men weren't reliable.

OVER THE NEXT YEAR, EMILY'S LETTERS AND PHONE CALLS became less cheerful. She complained of headaches; the job had become more tedious; the boss was insufferable. Her sister had moved away after marrying a man from Fujian Province. Emily's coworkers came and went; she was still the oldest woman in her office. By now, she had adopted a protective role, guarding new women employees against the boss's advances. For Christmas, she sent me samples from her factory: bracelets made of pink and purple plastic beads. She told me that I could give them to my sisters in America.

I sent her a story that I had written about her and her classmates, and she responded, "I'm not confident that I'm as admirable as I seem to you. It's true that I like to be alone. But partly it's because I don't know how to join the people; I can't share their joys and sorrows and cares. Although I really wish to." Another time she wrote about her job, "My head aches sometimes. And mistakes often happen. . . . Do you know of any kind of jobs that are interesting and benefit the whole society?"

I encouraged her to find something that required English skills, but I knew that my advice wouldn't help much. There was something elusive about her unhappiness—I had glimpsed a similar quality in some of my brightest students, most of whom were girls. The best freshman student in the English department had been a quiet girl who kept apart from her peers. After class, she often came to see the foreign teachers for extra practice with her English, which was excellent. During summer vacation, she returned to her hometown and killed herself by jumping off a bridge. I never learned anything else about her death; nobody in the class had been close to her. In China, more women killed themselves than men, and the suicide rate for females was nearly five times the world average—the highest of any country in the world. Many of them were women from rural areas who were in the process of becoming urban, middle-class citizens. There was something traumatic about this transition, even when it seemed to involve a better material life in places like Shenzhen.

One of Shenzhen's economic innovations was the establishment of "talent markets," or employment centers, which replaced the Communist-style system in which jobs were assigned by the government. Although these markets encouraged independence, they also subscribed to traditional notions of how a woman should look. Emily complained that prospective employers often thought that she was too short. She stood 1.53 meters tall—just over five feet—and the talent markets generally advertised jobs that required women to be at least 1.6 meters tall, especially if they wanted to be employed as receptionists, secretaries, or waitresses at expensive restaurants. Emily also said that her small eyes and big lips made it harder to find jobs that required female applicants to be blessed with *wuguan duanzheng*, meaning that "the five senses"—the ears, the eyes, the lips, the nose, and the tongue—were regular. In the first letter she wrote to me after starting work at the jewelry factory, she introduced her coworkers by carefully describing their appearance:

> Lijia is the most beautiful, able and shortest girl, who is liked by everyone. Huahui is a classical beauty, most private telephone calls from boys are for her. But I don't like her very much, for her word sometimes hurtful. Lily is the other secretary, who came two days earlier than me. She leaves us an impression of stupid and irresponsible. So she is not very popular in the office. Xinghao is the fattest girl concerning much about losing weight.

Although conventional notions of beauty could make things uncomfortable for women in the Shenzhen job market, they also enjoyed a degree of freedom in the city that wasn't possible for women in other parts of the mainland. In Shenzhen, it wasn't uncommon for young people to live together before marriage. Divorce was more acceptable, and many women were in no hurry to get married. During my wanderings around the city, I picked up two copies of a women's magazine entitled *Window to the Special Economic Zone*. Its articles carried such headlines as "One Night's Love," "I Am Not a Lady," "A Trap Set by an Old Man," and "Why Have an Abortion?"

Whenever Emily talked about difficult personal issues, she mentioned a radio call-in show, *At Night You're Not Lonely*, and its host, Hu Xiaomei, who was an idol to the young women of Shenzhen. In 1992, when she was twenty, Hu Xiaomei had left her home in a remote coal-mining region of the Chinese interior and moved to Shenzhen, where she found a seventy-dollar-a-month job in a mineral-water plant. One night, she telephoned a call-in show. Unlike most callers, she didn't want advice—she just wanted to tell listeners about her long-held dream of becoming a radio-show host. Hu Xiaomei was a good talker, and after she was finished she gave the listeners her work address and phone number.

"The next week, I got stacks of letters and more than a hundred phone calls," Hu Xiaomei told me when I met her during one of my trips to Shenzhen. "But the mineral-water plant fired me for using their phone for personal reasons, so I had no job. I took all the letters, bundled them together, and brought them to the radio station."

Hu Xiaomei paused and took a long drag on her Capri menthol superslim. We were sitting in a private room in a downtown Shenzhen restaurant. A pretty, petite woman with small features and long black hair, she was the type of Chinese smoker who could eat and smoke simultaneously. She exhaled a thin menthol stream and continued: "They said I was too young— I was only twenty years old and I had no experience. But one official decided to give me a chance. I told him that I was only twenty and that I didn't understand many things, but that many listeners were the same as me, so maybe I'd understand them."

A decade later, a million people tuned in to Hu Xiaomei's show every weeknight, and many of them were young factory women who worked outside the fence. Even though the Shenzhen radio station was government-owned, like all media in China, Hu Xiaomei often gave blunt advice that rankled tradition-minded officials—for example, recommending that young people not be afraid of living together before marriage. When I met her, she had written one book and was working on another. She often appeared on television, and despite her lack of formal education, she had an air of sophistication. When I asked her which writers she admired, she mentioned the stories of Raymond Carver. ("You can tell so much from a very small detail.") Like many women I met in Shenzhen, she was worldly, self-made, and confident. She had once broken off a long-term romantic relationship partly because, she said, "he didn't like it that people knew him as Hu Xiaomei's boyfriend."

All the women at Emily's factory listened to Hu Xiaomei's show religiously, and the next day they discussed the callers whom they had heard the night before: the wives who were having affairs, the mistresses who wanted a way out, the women who couldn't decide whether or not to move in with their boyfriends. Emily was most impressed by the individuality of Hu Xiaomei's responses. "She doesn't make blanket judgments," Emily told me. "She looks at each caller's specific situation and then decides."

But even with the help of Hu Xiaomei, Emily didn't find it easy to adapt to Shenzhen's freedom. She said that she thought it was acceptable for young people to live together before marriage but not to tell anybody about it. Hu Xiaomei had once told a caller the same thing, and Emily agreed that this decision should remain completely unspoken. "It could influence the way people see you, especially if you break up later," she said. "It's better if you just don't say anything about it." One day, when she remarked that sex was more open in Shenzhen than in other parts of China, I asked her whether this was good or bad. She thought for a moment and then said, "It's better than it was in the past. But it shouldn't cross a certain line."

"What line?"

"It has to do with traditional morality."

I asked her what she meant, and she rested her chin on her hand, thinking hard. "Traditional morality," she said again. But she couldn't define what it meant.

ONE DAY, I GAVE EMILY A COPY OF A POPULAR SHENZHEN novel, *You Can't Control My Life*. The book, which was published in 1998, follows the fortunes of its migrant heroine in Shenzhen from her first job as a secretary to a life of luxury as the mistress of a wealthy Hong Kong businessman. The author, a twenty-nine-year-old woman named Miao Yong, grew up in remote Gansu Province, in the far west, and migrated to Shenzhen after graduating from a teachers college. She found a job as a secretary, wrote fiction on the side, and became rich when *You Can't Control My Life*, her first novel, became a best-seller. The government banned the book because of its portrayal of drugs, gambling, and casual sex. Like many book bannings in China, this actually boosted sales—although all of the copies were bootlegs. Vendors sold black-market versions all around downtown Shenzhen. In front of the Stock Exchange, I saw one man selling *You Can't Control My Life* alongside Chinese translations of *Mein Kampf*.

"When I say 'you,' I mean society," Miao Yong told me when I asked about her book's title. "I'm saying that my life is controlled by me; it's not something for other people to take charge of." She explained that materialism was a key force in the novel. "Everything has to do with money; it's the first thing for everybody. In Shenzhen, it's always a question of exchange—you can exchange love for money, sex for money, emotion for money."

I met Miao Yong at a trendy Western-style café near her apartment in a Shenzhen luxury high-rise. She chain-smoked Capri menthol superslims throughout our meal, and she explained that her writing had been influenced by the novels of Henry Miller. ("His books were banned, too.") Despite the ban and the bootlegs, writing had made Miao Yong rich, because she had turned her novel into a popular television series. In order to get past the censors, she had purged the book of its most sensitive material, and she had also given it a happier title, for good measure. In the future she planned to be less explicit about the fact that Shenzhen was her setting. She believed that the cadres had banned her book because it gave the experimental city a bad name.

The first detail on the author's bio of Miao Yong's book was her blood type. Like many hip young Chinese, she believed that blood type helps determine character. She was twenty-nine years old, and she had recently purchased her first automobile. The politically correct title of her television series was *There's No Winter Here*. Miao Yong was type O. She told me that individualism was what interested her the most about Shenzhen. "In the past, China was very collective," she said. "It was all about group thought. But now, in places like Shenzhen, you can decide exactly what kind of person you want to be."

Self-invention was a core principle of Shenzhen, and some figures had become legendary because of their transformation: the girl from a coal-mining town who became a radio star, the secretary from Gansu who struck it rich from writing. Private lessons in subjects like English were popular, because people hoped to rise in the factory world, and there were also plenty of shortcuts. In front of the local Walmart, vendors sold bogus bachelor's degrees and transcripts for less than $100. Some migrants found other illegal ways to make money fast. I interviewed one Sichuanese prostitute who was hoping to limit the job to an eight-month run; she was making $740 a month, and she figured that soon she'd have enough saved up to return home and start a small business before anybody knew what she'd been up to. She had been a virgin when she first arrived in Shenzhen. She was twenty years old and obsessed with the notion of getting her old life back. I talked with another young woman who was working as a "three-accompany girl," a nightclub job in which women accompanied men as they eat, drink, and sing karaoke. It was a notoriously vague profession, and many three-accompany girls were willing to perform additional services. The woman I met claimed that she hadn't accepted money for sex, but she spoke nostalgically about her old job at a shoe factory, where she had earned $100 a month and slept in a dorm room with seven other workers. "My heart was more open in those days," she said. Now she earned $35 a night and slept all day; she had lost contact with all of her old friends. On her evenings off she liked to get drunk and go dancing in night clubs, entirely by herself.

After reading Miao Yong's novel, Emily said that she felt no connection with the world it depicted. The heroine had no heart—all she cared about was money—and she went from one man's bed to another. "It's too chaotic," Emily said. "You need to control this part of your life."

I asked her where she thought these new notions of morality had come from. She shrugged. "Most people say that they came from the West, after Reform and Opening. I think that there's probably some truth to this."

"What do you think the book's philosophy is?"

"It's saying that Shenzhen is a new city without any soul," she replied. "Everybody in the book is in turmoil—they can't find calmness."

AFTER DATING FOR A YEAR, EMILY AND ZHU YUNFENG started living together. They rented a three-room apartment in a small factory town about thirty miles beyond the Shenzhen fence, near the household appliance plant where Zhu Yunfeng worked. The apartment building was closer to being finished than most of the neighboring structures. The concrete stairways were cracked, but everything worked, and the kitchen was well equipped. It was the first decent home that Emily had had since coming to Shenzhen.

Another young Sichuanese couple also lived in the apartment. Each of the couples had their own bedroom, but they shared the living room, which was furnished with a color television, a VCR, a low table, and a bed that doubled as a couch. All four of them got along well. One of the bedrooms had a laminated poster of a topless foreign couple making out. Such posters were popular in China; they were considered romantic, and they were inoffensive because they portrayed a non-Chinese couple.

Back home in Fuling, Emily never would have moved in with a man before marriage, and she didn't tell her parents about the apartment. But one day, during a phone conversation, her mother asked if she and Zhu Yunfeng were living together. "I didn't say anything," Emily told me. "She knew from my silence that it was true." After that, neither of them said anything more about it.

Emily still spent weekday nights in the jewelry-factory dormitory, but on weekends she stayed at the new apartment. Zhu Yunfeng had been promoted to a supervisory position in his factory, and he was earning $360 a month, an excellent wage. Emily's salary had risen to $240 a month, but she was sick of the job. She didn't like the way the boss occasionally asked her to work overtime, and she didn't like being restricted to the dormitory at night.

One weekday evening, Emily broke curfew and spent the night with Zhu Yunfeng. The next morning, the boss called her into his office.

"He asked me what time I came back last night," she said.

"That's the way he was—it was never direct. He didn't ask me whether or not I had come back—he just asked what time. I said, 'I came back this morning.' I didn't make any excuse or explanation. He didn't know what to say. I don't think he knew whether to get angry or laugh. He looked at me, and finally he just walked away."

A few weeks later, another young woman at the factory started breaking curfew.

Not long after that, the boss took a pretty worker off the production floor and made her his personal secretary. The worker was from Hunan Province, and she was eighteen years old. Emily spent a lot of time telling the girl stories about the boss, and one day he confronted Emily. He asked her what people said about him behind his back, and then he finally got to the point.

"Do you tell the other workers that I'm lecherous?" he said.

Emily said, "Yes."

He tried to laugh it off, but it was clear that he no longer liked having her around. Emily started spending her free time searching for work. A few weeks later, she found a position as a nursery-school teacher. Like the jewelry factory, the school was outside the fence, but there were no Taiwanese bosses, no factory dormitories, and no evening shifts. The salary was roughly the same as what Emily had been making at the plant. She was going to teach English.

When she told her boss that she was quitting, he tried to give her a dressing-down.

"You've changed," he said. "You used to be obedient. Everything changed after you got a boyfriend."

"I didn't change," Emily said. "I just got to know you better."

On my last night in Shenzhen, Emily and I went outside to listen to the radio. She thought it was best if we left the apartment, because Zhu Yunfeng had come home late and he was in a bad mood. It was a warm, clear night and the stars were bright above the dormitory lights of the factories.

Earlier that day, a young man had been injured while working under Zhu Yunfeng's supervision at the factory. Zhu Yunfeng didn't say much about it to Emily, other than that he wanted to be alone. The factory had been working overtime in hopes of meeting orders for a new product line, and the combination of time pressure and unfamiliar production equipment was always bad for accidents. The new product was a metal thermos bottle.

Emily and I climbed a nearby hill where we could look out over the town. It was a typical factory settlement beyond the Shenzhen fence: a cluster of shops and apartment buildings wedged into a dusty cut in the hills, and then, fanning out along the two main roads, long strips of factories and their dormitories. There were shoe plants, clothes factories, and a computer-accessory factory whose top story had been gutted in a recent fire. The smoke had left the plant's white tile walls streaked with black. Emily said that nobody had been hurt in the fire, but down the road there was another factory where several workers had died a few years ago in a massive blaze. That factory had been producing Christmas decorations and lawn furniture.

Emily was about to start her teaching job, and she was a little apprehensive. She worried about her English, which had slipped during the years at the factory, and she wondered whether she would be able to discipline the children. But she liked the school campus, and she smiled whenever she talked about the new job. She kept her hair short now, her bangs pinned back with plastic barrettes. Around her neck she wore a simple necklace that Zhu Yunfeng had given her—a jade dragon, her birth sign.

We sat together on top of the hill and listened to *At Night You're Not Lonely*. Neither of us said anything for a long time. The volume control on the radio was broken, and the voice of Hu Xiaomei crackled thinly in the night air. It was after eleven o'clock, and in the distance we could see one of the factory dormitories, a huge block of lighted windows on the horizon. The first caller started crying because she regretted the way she'd treated an old boyfriend, who had left her. Hu Xiaomei told her that the experience would be good for her and that maybe the next time she'd get it right. The second caller talked about missing his high-school girlfriend, who was far away, working in another part of the country. Hu Xiaomei said that he shouldn't believe that every feeling was love. She told the third caller that, at twenty-three, she was wrong to feel as though she had to get married immediately.

Down below, the lights in the dorm were being turned off, one by one. I thought about the worker who had been injured earlier that day, and I remembered a recent conversation with Emily. We had been talking about how migrants handled the new personal freedoms in Shenzhen, and Emily said that she admired the way that people learned to help themselves. She had often made this comment in the past, but now she added that sometimes the isolation also frightened her—all these people living on their own. "In original society," she said, "people lived in groups. Eventually, these groups broke down into families, and now they're breaking down again, into so many different people. Finally, it will be just one single person." A few days earlier, she had remarked that the changes in Shenzhen—the independence of young people, the shift from control by the Communist Party to control by the factory bosses—had come too abruptly. "If you could have some kind of perfect socialism, that would be the best," she said. "But it's impossible. That was just a beautiful ideal."

Now, sitting on the hillside, I asked her if she wanted to leave Shenzhen. She shook her head quickly. I asked how she thought the new pressures of the city would change the people who lived here.

"The result is that people will have more ability," she said. "And they'll have more creativity. Afterward, there will be more different ideas. It won't be a matter of everybody having the same opinion."

I asked, "How do you think this will change China?"

She fell silent. In the distance, most of the dormitory lights had flickered out. I had no idea how I would have answered the question myself, although I liked to think that once people learned to take care of themselves, the system would change naturally. Still, I had seen Shenzhen's fragmentation—the walled city, the locked factories, the people on their own, far from home—and I wondered how all of it could ever be brought together into something coherent.

I looked at Emily and realized that the question wasn't important to her. Since coming to Shenzhen, she had found a job, left it, and found another job. She had fallen in love and broken curfew. She had sent a death threat to a factory owner, and she had stood up to her boss. She was twenty-four years old. She was doing fine.

# UNDERWATER

JUNE 7, 2003

AT 6:13 IN THE EVENING, AFTER THE ZHOU FAMILY HAS AL-
ready moved their television, a desk, two tables, and five chairs
onto a pumpkin patch beside the road, I prop a brick upright
at the water's edge. On new maps for the city of Wushan, this
body of water is called Emerald Drop Lake. But the maps were
published before the lake existed, and the color is turning out
to be a murky brown rather than emerald. The lake itself is actually an inlet of the Yangtze River, which for the past week
has been rising behind the Three Gorges Dam. On Zhou Ji'en's
next trip down from his family's bamboo-frame shack, he carries a wooden cupboard on his back. He is a small man with a
big smile, a pretty wife, and two young daughters; and until
recently all of them were residents of Longmen Village. The village does not appear on the new maps. A friend of Zhou's carries the next load, which includes the family's battery-powered
clock. The clock, like my wristwatch, reads nearly 6:35. The
water has climbed two inches up the brick.

Watching the river rise is like tracking the progress of the clock's short hand: it's all but imperceptible. There is no visible current, no sound of rushing water—but at the end of every hour another half foot has been gained. The movement seems to come from deep within the river, and it alarms every living thing on the shrinking banks. Beetles, ants, and centipedes radiate out in swarms from the river's edge. After the water has surrounded the brick, a clump of insects crawl madly onto the dry tip, trying desperately to escape as their tiny island is consumed. Most residents of Longmen left last year, when the government relocated them to Guangdong Province, in the south of China. But a few, like Zhou Ji'en and his family, stayed behind to work the land for one last spring. They knew that the flood was coming, but they had no idea that it would move so fast. Two days ago, the elder daughter, Zhou Shurong, completed first grade. Yesterday, her mother, Ou Yunzhen, harvested the last of their water spinach. Today those plots are underwater. All that remain are pumpkins, eggplants, and red peppers.

More than thirty feet above Zhou's pumpkin patch, a neighbor named Huang Zongming is building a fishing boat. Huang has told me that it will be "two or three more days" before the river reaches the boat. Chinese farmers tend to speak about time in an indeterminate manner, even at a moment like this, when the river is a very determined two and a half days ahead of schedule. The government says that it will ultimately rise by more than two hundred feet.

The Zhou family has rented a three-room apartment on the hillside above, and they are moving their belongings today because several inches of water have crept across the only road out. At 7:08 P.M., the brick is half submerged. Zhou Shurong has carried out her possessions—an umbrella, an inflated inner tube, and a Mashimaro backpack that contains a pencil box and schoolbooks. As the adults continue hauling furniture, the little girl sits at a table in the pumpkin patch and calmly copies a lesson:

*The spring rain falls softly,*
*Everybody comes to look at the peach blossoms.*

At 7:20, a young man arrives on a motorcycle and catches black scorpions that are fleeing the rising water. "Usually they're hard to find," he tells me. "They're poisonous, but they can be used for medicine. In Hunan Province, I've seen them sold for a hundred yuan a pound."

By 7:55, when a flatbed truck appears with two moving men, the brick has vanished. Only a small stretch of road remains dry, and that's where they park the truck. The river hits the vehicle's front left tire at 8:07 P.M. The men start to load the furniture; the sky grows dark. "Hurry up!" Ou Yunzhen says. "If you don't, the truck won't be able to get out!" Nearby, Zhou Shurong and her five-year-old sister, Zhou Yu, stand the way children do when they sense anxiety in adults: perfectly still, eyes unblinking, arms straight at their sides. At 8:23, the water reaches the left rear tire. The television is the last object to be loaded; it placed with great solicitude on the front seat, right next to the girls. At 8:34, the driver turns the ignition. The water reaches the top of the hubcaps as the truck rumbles off. After it's gone, Ou Yunzhen stays behind to harvest the last of the pepper crop in the dark.

The next day, in the hard light of the afternoon, I visit the remains of the shack to see what the family abandoned to Emerald Drop Lake. There's a man's left boot, a smashed metal flashlight, half of a broken Ping-Pong paddle, an empty box that says, in English, "Ladies Socks," and a math examination paper, written in a young girl's hand, with the score at the top in bright red ink: 62 percent.

FROM 1996 TO 1998, I TAUGHT ENGLISH TO COLLEGE STUDENTS in Fuling, a small city on the Yangtze some two hundred miles upstream from Wushan. Every winter, along with the other seasonal changes, the river shrank. Rain became less frequent, and the snowmelt stopped coming off the western mountains, until eventually the Yangtze exposed a strip of sandstone known as the White Crane Ridge. The ridge was thin and white, and it was positioned parallel to the river's current, like a long low boat that had been moored offshore. It was covered with thousands of inscriptions—for centuries, local officials had used it to keep track of low-water levels. When I visited the ridge in January of 1998, the river was exactly two inches lower than it had been at the time of the earliest dated carving, 763 A.D. The inscriptions made it clear that, in these parts, the Yangtze's own cycle mattered more than the schedules of any authority. One carving, completed in 1086 A.D., commemorated the ridge's emergence during the ninth year of the reign of Yuanfeng, an emperor of the Northern Song dynasty. In fact, Yuanfeng had died the year before, but news of the death—and of the new emperor—had yet to reach the river.

Fuling was still remote when I lived there. The city had no traffic lights, no highway, and no railroad. There was one escalator in town, and people concentrated hard before stepping onto it. The only fast-food restaurant bore the mysterious name California Beef Noodle King U.S.A. It was a poor region, and it got poorer as you went downstream toward Wushan. Wushan was at the center of the Three Gorges—a 120-mile stretch of narrow riverway, bordered by high mountains and steep cliffs, where the landscape was known for its stunning beauty, hard farming, and fast currents.

I could see the river from the classroom where I taught writing. Our government-issued textbook included a unit on "Argumentation," which featured an essay titled "The Three Gorges Project Is Beneficial." The essay cited a few drawbacks—lost scenery, displaced people, flooded cultural relics—but the author went on to assure that these were easily outweighed by the benefits of better flood control, a greater supply of electricity, and improved river transportation. Given the fact that the Chinese government had strictly limited public criticism of the dam, we could go only so far with the Argumentation unit. I spent a lot more time teaching the proper form of an American business letter.

The idea of building a dam in the Three Gorges had been around for almost a century. Sun Yat-sen proposed it in 1919, and after his death the vision was kept alive by dictators and revolutionaries, occupiers and developers. Chiang Kai-shek promoted the idea, as did Mao Zedong. The Japanese surveyed sites during their occupation in the 1940s. Engineers from the U.S. Bureau of Reclamation helped the Kuomintang; Soviet technicians advised the Communists. But by the time construction finally began, in 1994, the era of big dams had passed in most parts of the world. Both the U.S. government and the World Bank refused to support the project, because of environmental concerns. Many critics of the dam believed that one of its main goals—protection against the floods that periodically ravage central China—would be better served by the construction of a series of smaller dams on the Yangtze's tributaries. Engineers worried that the Yangtze's heavy silt might back up behind the Three Gorges Dam, limiting efficiency. Social costs were high: More than a million people had to be resettled, and low-lying cities and towns would be rebuilt on higher ground. Once completed, the dam would be the largest in the world—as high as a sixty-story building and as wide as five Hoover Dams. The official price tag was more than $21 billion, roughly half of which would be funded by a tax on electricity across China.

But in Fuling I never heard any of this. When I left the city, in the summer of 1998, the only visible indications of the project were a few altitude markers that had been painted onto buildings in the lower sections of town. The signs said, in bright-red paint, "177 M"—the future height of the reservoir. It was precisely forty meters higher than the White Crane Ridge's inscription from 1086 A.D. During the next five years, I often returned to the Three Gorges, where more red signs cropped up along the river. Most identified meter marks of either 135 or 175, because the reservoir was scheduled to be filled in stages, first in 2003 and then at the higher level in 2009. But other numbers were also common: 145, 146, 172. Some of the signs had an odd specificity: 141.9, 143.2, 146.7. They reminded me of the ridge—the whole valley was being marked and inscribed in preparation for the flood.

Down near the riverbanks, the old cities and villages were left virtually unimproved. Even though the rest of China was in a construction frenzy, there was no point in building anything new where the water was certain to rise. For a time, these settlements gave a rare glimpse into the past: still-life portraits of gray brick and dark tile. Above them, there was usually a green stretch of empty hillside or cropland punctuated by the occasional red sign. And then, high above the level of the future reservoir, new cities were being built with cement and white tile. These horizontal bands were like the strata of a geologist, except that they looked to the future as well as the past. You could see it all at a glance: the dark line of riverside settlements, the green stretch of farmland that would be claimed by the reservoir, and the clusters of white looking toward tomorrow.

The new cities followed distinct stages of creation. In the beginning, there were mostly men: construction workers, bulldozer operators, dump-truck drivers. Shops soon appeared, but they stocked almost nothing that you could eat, drink, or wear. The necessities of this place were different: tools, windows, lighting fixtures, bathroom fittings. Once, in the new city of Fengdu, I walked down a half-built street where almost every shop was selling doors. Lamps and sockets were sold long before there was reliable electricity. I visited areas where the roads were dirt and people used pit toilets, but their stores stocked everything needed to furnish a modern bathroom. It was always a good sign when women appeared in the new towns—this meant that the construction had moved beyond the point of basic infrastructure. When you saw children, you knew that the new city was alive.

The process of demolition was more erratic. The government began tearing down old towns in 2002, and most residents were given housing compensation, which they could use to purchase apartments in the new cities. But an estimated hundred thousand rural villagers were relocated to other parts of China. Generally, they were moved in blocs; sometimes an entire hamlet was loaded onto a boat and sent downstream to another province, where the government provided small land allowances. I knew a cop who escorted an entire village by train to the southern province of Guangdong. He rode with the villagers for two days, walked them out of the Guangdong station to waiting buses, and then turned around and got back on the train.

In the final stage of destruction, a town's shops sold nothing but products that people could eat, drink, or wear. It was common to see elderly people—some of them couldn't bear to leave, and others had no children or relatives to help them move. The young people who stuck around were usually trying to find some last profit in their community. Scavengers ripped out the scrap metal from buildings, and farmers tried to coax a final crop out of the doomed soil. Neat rows of vegetables were cultivated amid the rubble, like gardens in a war zone. In the village of Dachang, I arrived just as the first row of houses was being torn down. A middle-aged man sat in the wooden frame of his ruined home, drinking grain alcohol. It was ten o'clock in the morning and he was quite drunk. "I'm like a man hanging from a nail," he told me.

Some of the stragglers were like that—they had slipped outside the valley's evolution. Often, they didn't belong to traditional Communist work units, or they were farmers who hadn't had much land, or they were residents who had been registered

in another part of China, which meant that they didn't qualify for compensation. When I visited the old city of Wushan
in September of 2002, after most of the structures had been
torn down, there were still a number of beauty parlors where
prostitutes waited patiently behind blue-tinted windows. I had
a sudden vision of the women, nine months later, sitting with
the water up to their necks. In the tiny village of Daxi, one old
man pulled out form after form, showing me how he had lost
his resettlement funds—about twelve thousand dollars—on a
bad investment in coal. In Dachang, which had the most intact
Ming and Qing dynasty architecture in the region, I was shown
around by a man in his early twenties named Huang Jun. At the
old dock, underneath a massive banyan tree, he pointed out two
stone lions that guarded the steps leading down to the water.
The lions' faces were chipped and scarred; their backs had been
worn smooth from decades of being sat on. In the future, this
site would be underwater.

"During the Cultural Revolution, Red Guards threw the lions into the river," Huang said. "It was chaotic, and nobody
knew what had happened to the statues. But years later an
old man dreamed one night that they were out in the water.
He told the other villagers, and they searched in the river and
found them. That was in 1982—I remember it. The story is very
strange, but it's true."

JUNE 8, 2003

AT 9:40 IN THE MORNING, THE RIVER IS ALMOST EMPTY.
Tourist ships have been canceled for weeks—first because of the
outbreak of SARS, and more recently because the reservoir has
been filling up. Today, the water continues to rise at roughly six
inches an hour. Some friends and I are in a small boat, powered
by an outboard motor, and we head in the direction that used

to be downstream. But now the river has no heart; it sits dead between the walls of the Wu Gorge. The water comes to life only in the bends, where the sky opens and the wind kicks up a few waves.

Eight months ago, I hiked this route on a series of century-old footpaths. The trails had been carved directly into the limestone cliffs, and they were suspended more than two hundred feet above the river. I traveled with a friend, camping along the way, and in the middle of the gorge we followed a tributary called the Shennu Stream. The Shennu came from the southern mountains and it was too shallow for boat traffic. Hiking upstream, we watched the valley grow deeper, until at last we were in a deep canyon, hopping from boulder to boulder. Ferns hung high in the cliffs; there weren't any red signs in this unpopulated place. We stood in a shaft of sunlight, trying to guess how high the cliffs would be filled by the new reservoir.

This morning, I ask the boat pilot to head toward the Shennu. We move east through the Wu Gorge, where I look for the cliff trail, but virtually all of it is already underwater. The mouth of the Shennu is wide and still; tree branches float near the banks. But upstream, where the water is moving, the debris begins to disappear, and the river's color starts to shift. It turns dark green, then green-blue. Long-stemmed ferns dangle directly into the river—the current hasn't yet pulled them off the cliffs. We zip around one sharp bend, then another. This gorge still has the character of a waterway carved by a smaller stream—it meanders wildly, taking abrupt turns—and it's mesmerizing to flash through it on a fast boat. The boat doesn't belong here; the water doesn't belong here: this is a day-old gorge. The stream color shifts to blue, and the sound of rapids rises above the motor. Nobody sees the rock until we hit it.

There is a horrible scraping noise and the boat lurches to a stop; all of the passengers grab the side. The engine cuts off. We sit in stunned silence for a moment while the pilot checks for damage. Drifting backward, the rapids suddenly loud now, we see the boulder, its tip glistening beneath a foot and a half of water. Somebody remarks that by tomorrow it will be deep enough for safe passage. Looking at the submerged shape—smooth and rounded, like some hunchbacked animal hiding in the shallows—I think of the stone lions and the old man's dream.

AT 2:50 THAT AFTERNOON, I RETURN TO EMERALD DROP LAKE. Yesterday, I walked in on the road; today I have to take a boat. I visit the Zhous' former residence with Huang Po, the nine-year-old boy from the fishing family next door. When Huang sees the 62 percent math examination lying in the rubble, he picks it up, folds the paper carefully, and puts it into his pocket.

"Why do you want that?" I ask.

"If I see her, I'll give it to her," Huang Po says.

"I don't think she wants it," I say.

The boy grins slyly and touches his pocket.

His father, Huang Zongming, has finished building the fishing boat a day earlier than he predicted. It is forty-two feet long, made of the wood of the Chinese toon tree, whose boards have been hand-cut and joined with heavy iron rivets. The labor required several family members and more than twenty days of work; recently they caulked the cracks with a mixture of lime, hemp, and tung oil. This morning, they treated the entire surface with another coat of the oil. Tung oil is a natural sealant, as well as an effective paint thinner—in the 1930s, it was China's single most valuable trade product. The oil gives the wood a reddish shine; the shape has a rough, simple beauty. The boat sits on wooden struts. It has never touched water. I ask Huang when he plans to move it.

"Whenever the water reaches it," he says. Huang is shirtless, a skinny, square-jawed man with efficient ropelike muscles. Later, when I ask if he's worried about the boat's not being tested before the water rises, he gives me the slightly annoyed look of a shipwright hassled by diluvian reporters. Huang Zongming is a righteous man, and he knows that his boat will float.

I first met the Huang family in September of last year, when Longmen Village was still on the map. By local standards, the place was relatively prosperous; residents fished the waters of the Daning River, the tributary that enters the Yangtze at Wushan, and they farmed the rich floodplain. When most of the villagers were relocated to Guangdong, Huang Zongming and his brothers, Zongguo and Zongde, stayed behind. The government had organized the Longmen transfer, but it was unable to ensure that everybody actually went. In today's China, despite official registrations, people who are determined to live in a place can generally find a way to do so.

Huang Zongguo told me in September that the relocated villagers complained about the scarcity of farmland in Guangdong. He added that they struggled because they couldn't understand Cantonese. Huang was disappointed with his family's resettlement allowance—roughly ten thousand yuan per person, or a little more than $1200. We spoke in his simple brick home, where the electricity and water had just been cut off. Outside, the village was eerily quiet. In two weeks, Huang Zongguo said, the demolition crews would arrive.

Since that visit last fall, Zongguo and Zongming have paid for the construction of a two-story building high above Emerald Drop Lake. The home is almost completed, but Zongming tells me that he doesn't want to move there until later in the year; during the summer he prefers to be close to the water. He lives in a shack topped by an old fiberglass boat cover, along with his wife, Chen Cihuang, and their two children, Huang Po and Huang Dan, a twelve-year-old girl. Zongming is thirty-five years old, and he has worked a fishing boat since he was ten. Unlike his former neighbors, the Zhous, who grew up in the highlands and migrated to Longmen to farm, Zongming is completely at ease around water. His response to the rising river is simply to move the makeshift home thirty feet higher up the hillside. But he feels no need to undertake this project until the last possible moment. At 5:15 in the afternoon, the water is about eight feet below the shack. I ask Zongming when he expects the river to reach his new boat. "Probably about noon tomorrow," he says.

IN THE NEW CITY OF WUSHAN, WHICH HAS BEEN CONSTRUCTED on the hillside directly above the former site, there are parallel streets called Smooth Lake Road and Guangdong Road. Smooth Lake refers to a poem that Mao Zedong wrote in 1956, when he swam across the Yangtze and dreamed of a new dam:

> Walls of stone will stand upstream to the west
> To hold back Wushan's clouds and rain
> Till a smooth lake rises in the narrow gorges.

The poem is familiar to people along the river, and they often speak of the project in terms of national welfare and achievement. In 1997, when the Yangtze was diverted at the dam site to prepare for construction, President Jiang Zemin proclaimed, "It vividly proves once again that socialism is superior in organizing people to do big jobs." Once, in the new village of Qingshi, I saw a restaurant whose owner had posted a handwritten sign in the form of a traditional Chinese New Year's couplet:

> Honorable migrants leaving the old home
>     and moving to a new life
>
> Giving up the small home, serving the nation,
>     building a new home.

Guangdong Road takes its name from the southern province that was the first part of China to boom after free-market reforms were announced in 1978. Funding from this area has supported part of the Three Gorges Dam development, and names in the new towns often pay respect to the south. Wushan has a school called the Shenzhen Bao'an Hope Middle School. Shenzhen is the thriving Special Economic Zone that helped drive Guangdong's development, and seeing its name on a school in Wushan is like coming upon Silicon Valley High in the wilds of Appalachia. New Wushan itself seems like a vision of prosperity that has arrived from far away. The town's central square features an enormous television screen, and crowds gather there at night to watch martial arts films. Guangdong Road is lined with plastic palm trees that are illuminated after dark. There is a fake Starbucks. Other shops have English names: Well-Off Restaurant, Gold Haircut, Current Bathroom. There's a clothing boutique called Sanity.

In the new cities I rarely hear criticism of the dam. Even in rural areas, where people have received far fewer benefits, complaints tend to be mild and personal. Generally, people feel that they haven't received sufficient resettlement allowances, which they blame on the corruption of local Communist Party cadres. But those who complain almost never question the basic idea of the dam. When I asked Huang Zongming what he hoped his children would do after they became adults, he said he didn't care, as long as they used their education and didn't fish. He told me that the dam was good because it would bring more electricity to the nation. In Wushan a cabbie told me that his hometown had leapfrogged a half century. "If it weren't for the dam, it would take another fifty years for us to reach this stage," he said.

But later in the same conversation he told me that the city wouldn't last another half century, because of landslides. The new Wushan, which has a densely concentrated downtown population of fifty thousand, is a vertical city: high buildings on steep hillsides that have never been heavily settled. Concrete erosion controls prop up many of the neighborhoods. The cabbie drove me to Jintan Road, where there had already been a landslide. An apartment building had been evacuated; piles of dirt still pressed against the street. I asked the cabbie if he was concerned about the fifty-year limit. "Why worry about it?" he said. "I'll be eighty by then!"

During my years along the Yangtze, I had always been impressed by the resourcefulness of the people, who responded quickly to any change in their surroundings. They took the revolution of the market economy in stride; if a product was in demand, shops immediately stocked it. You could find people doing business anywhere, even at both extremes of the resettlement process. That was one thing that connected both the dying villages and the brand-new cities: somebody always found a way to sell what was needed, whether it was bathroom fittings or instant noodles. But there was almost no long-term planning. If the river rose, they moved up the hillside; farmers waited until the water reached their fields before harvesting. When people spoke of the future, they meant tomorrow.

Once, I discussed this shortsightedness with Jiang Hong, a Chinese-born geographer who taught at the University of Wisconsin at Madison. She had studied communities in the deserts of northern China, where generations of government policies had been implemented to convert the region into arable land. Many of these practices were environmentally unsound, and local residents generally resisted them, because they knew what was good for farming. But she had noticed that in recent years there had been less opposition to such schemes, partly because the free-market reforms gave people more incentive to try to change their surroundings. In the past, government campaigns often touted some abstract goal, like the attempt to surpass the United States and Britain in steel production in the late 1950s. Such a target could inspire a farmer for only so long—but nowadays everybody wanted a better television set or refrigerator.

And the lack of political stability taught people to avoid long-term plans. "Since 1949, policy has changed so often," Jiang told me. "You never knew what would happen. In the 1980s, people saw the reforms as an opportunity. And you had to seize the opportunity, because it might not last."

Whenever I traveled along the Yangtze, I sensed that the dam's timing was perfect. Building the dam appealed to the dreams of the Communist leaders, but they never could have achieved it in the days of Maoist isolation and political chaos, before the market reforms. And if the reforms had been around long enough for locals to get their bearings and look beyond satisfying today's immediate desires, they would have questioned and possibly resisted the project. In the future, when people look back at this particular moment in China's transition, with its unusual combination of communism and capitalism, the most lasting monument may well be an enormous expanse of dead water in central China.

JUNE 9, 2003

AT 9:30 IN THE MORNING, AFTER HUANG ZONGMING HAS drunk a single glass of grain alcohol, the waters of Emerald Drop Lake reach a corner of the wooden frame that is propping up the fishing boat. Most of the family's belongings have yet to be carried up the hill. Offshore, a snake glides through the reservoir, his head up like a periscope.

A resident of Wushan has commissioned Huang and his nephew to caulk another long rowboat, and Chen Cihuang has been left to organize the evacuation. She is dressed in a Beijing 2008 Olympics T-shirt and a pair of trousers whose print is imitation Burberry. At 9:46, the water touches another corner of the boat frame. The family has loaded the craft with a few possessions: a power drill, a basket of fishing gear, a carton of Magnificent Sound cigarettes. They throw spare lumber into the stern. An inch of water has flooded the former home; Chen, her sister-in-law, and the children wade through as they carry their belongings. Huang Po splashes his sister whenever possible.

By 10:47, Huang Po has made enough of a nuisance of himself to be excused from all moving duties. He strips off all of his clothes and goes swimming.

At 10:59, a sampan floats past and the pilot shouts out, "Do you have a boat to sell?"

Chen snaps back, "We're this busy and you think we're selling boats?"

By 11:40, all four corners of the frame are underwater. Sixteen minutes later, another sampan drifts past; somebody is selling coal. The move is complete when Huang Zongguo helps the women dismantle the old roof. Huang Po suns himself, stark naked, on the prow of the new boat.

At 1:34 in the afternoon, the wake of a passing craft rocks the fishing boat, which lurches, creaks, and finally swings free of the frame. It floats.

# THE URANIUM WIDOWS

THERE ARE MANY URANIUM WIDOWS IN SOUTHWESTERN Colorado, and some of them keep radioactive rocks around the house, but probably only one has a photograph of herself drilling for ore with a jackhammer while wearing nothing but sandals, denim shorts, and a bra. Her name is Pat Mann, and she is eighty-one years old. "You'll have to accuse my attire," she said with a laugh, as she handed me the photo. Mann explained that she dressed like that on hot days in the 1950s, when she worked on her first husband's mining crew. "It's a filthy job," she said. "People say, 'That uranium will kill you!' Well, we'd drill into those veins and blow it up, and it'd be all over. We'd be caked in it." I asked her about plans to build a mill nearby that would process uranium to be used for nuclear power generation. "I know we got a bunch of tree huggers and grass eaters," Mann said. "They seem to be against the mill. Most of them haven't lived with this stuff. I lived with it, and it hasn't bothered me."

Mann resides in a double-wide trailer in the remote town of Paradox. Around here, place names have the ring of parables: Calamity Mesa, Disappointment Creek, Starvation Point. The local history of uranium is long and often troubled, and the economy has been devastated since the Three Mile Island accident, in 1979, when Americans turned against nuclear power.

Many old-time Colorado miners suffer from lung disease, and one former mill community, Uravan, was deemed so radioactive that everything in town—houses, streets, even the trees—had to be shredded and buried. And yet since 2007, when a company called Energy Fuels arrived with plans to build America's first new uranium mill in almost thirty years, the response in the Paradox region has been overwhelmingly positive.

For outsiders, this reaction is puzzling. "Why would somebody want to go into something that killed people in horrible ways?" one newcomer asked. Environmental organizations have filed lawsuits to block the project, which would also lead to renewed mining, and they've expressed frustration with signs of growing national openness toward nuclear power. American nuclear plants still produce 20 percent of the nation's electricity, but a new reactor hasn't been licensed since 1996, and 86 percent of the uranium used for fuel is imported. Domestic mining and milling have been erratic since the mid-1990s, when the Megatons to Megawatts program was initiated, converting warheads from the former Soviet Union into fuel. But that program is set to expire in 2013, and the prospect of climate change has led to the re-evaluation of a power source that combines high yield with low carbon emissions. In 2010, President Obama approved more than $8 billion in conditional loan guarantees for the construction of new reactors. In former industry centers like southwestern Colorado, old debates have been rekindled. Some claim that cancer rates are the highest in the state; others say that this area is Colorado's healthiest. Natives told me that former mill sites are harmless, whereas environmentalists suggested rolling up the windows and driving by fast. I have yet to meet a uranium widow who opposes the industry that killed her husband.

Pat Mann has outlived two of them. The last one, George, died of lung cancer in 2000. "A lot of miners died from cancer, but they smoked," she said. "George was a heavy smoker." After we chatted for a while, Mann showed me her backyard rock collection, where she picked up a stone with yellow streaks so bright they could have been painted. She said that she didn't believe that uranium really causes cancer. She had big hands, with a mangled finger that had been surgically reattached years ago after it was crushed by a fifty-five-gallon oil drum. She put the uranium down, wiped her hands on her pants, and gave me a good firm shake before I left.

COLORADO'S ATOMIC HISTORY IS FULL OF CONTRADICTIONS, beginning with the fact that the first large-scale mill that processed radioactive elements was built in hopes of curing cancer. Around the turn of the twentieth century, when Marie and Pierre Curie began pioneering research on radioactivity, they worked primarily with radium. Soon it was being used in experimental treatments of cancerous tumors, a forerunner of radiation therapy. In the early 1900s, a Pittsburgh industrialist named Joseph M. Flanner lost his sister to cancer, and he blamed her death on a lack of radium. He responded with a tycoon's act of grief: in 1912, his company, Standard Chemical, built a mill to process ore less than ten miles from Paradox Valley.

Radium is exceedingly rare and highly radioactive. It's a decay product of uranium, and it gives off radon gas. Its peak price, in 1919, was more than $3 million an ounce—at the time, the most expensive substance on earth. Once, Marie Curie traveled all the way to the United States to receive a gram of Colorado-mined radium. But its value turned out to be fleeting; eventually it was replaced by more effective substances for radiation therapy and other applications. Coloradans switched to producing vanadium, another element found in local rocks, which can be used to harden steel. The old Standard Chemical mill was acquired and converted by Union Carbide, which built a town around the site.

They called the place Uravan in honor of the local elements. It was another parable of a name, although in the beginning they had no idea of the value of what they were sitting on. Until the 1940s, uranium had few commercial applications, and vanadium mills discarded the element in their tailings. During the Second World War, atomic-bomb scientists realized that those Colorado waste piles might help end the conflict. In 1943, the Manhattan Project built a new mill in Uravan, processing vanadium tailings into uranium oxide, or yellowcake. The yellowcake was sent to plants in other parts of the country, where it was enriched into bomb material, along with uranium from the Belgian Congo.

All of this was done in strict secrecy. The word "uranium" was banned from official reports, and workers didn't realize that they were contributing to the atomic bomb, at least until the attacks on Hiroshima and Nagasaki. ("It kind of leaked out after that," one elderly local told me in a slow drawl.) After the war, when the nuclear arms race heated up, the government encouraged private citizens to explore and drill for uranium. Federal agencies build backcountry roads, and the Atomic Energy Commission established guaranteed prices for ore. It became the only government-sponsored mineral rush in American history—approximately nine hundred mines were opened across the Colorado Plateau. In a region of tough names, these were the dreamers: the Hidden Splendor Mine, the King Solomon, the Silver Bell.

There was essentially no regulation. Most mines lacked proper ventilation, and in the 1950s public-health officials discovered that radon concentrations were nearly a thousand times higher than the accepted level of safety. Miners liked their cigarettes underground, where radioactive particles attached to the smoke and were drawn deep into the lungs. Meanwhile, Uravan's population grew to more than eight hundred, with its mill situated right in the center of town. One native told me that as a boy he descended into a uranium mine and ate lunch with his father on Bring Your Son to Work Day. If people heard that a mine was "hot"—highly radioactive—they couldn't wait to work there. In the nuclear cycle, a major risk of public contamination is mishandled mill tailings, which contain radium and other radon emitters. In Uravan, the sandy tailings served as bedding for construction projects. People laid water lines in the stuff. Gardeners used it to loosen up the clay soil. Old mill equipment, much of it contaminated, was tossed openly onto a hill above town; kids and scavengers liked the dump so much they called it Treasure Island.

WHEN A REMOTE COMMUNITY SUFFERS A HEALTH CRISIS, IT'S as if a curtain has separated people from the outside world. Typically, locals are frustrated that others fail to comprehend their pain, but the reaction in the uranium towns is the opposite. "He didn't take it real personal," Gayland Thompson told me when we discussed his father, a miner who died of lung cancer. "He wanted to work there." Like other people from the Uravan region, Gayland complained that protestors from elsewhere speak in the town's name, exaggerating health problems. From the locals' perspective, the curtain that separates them from the world also serves as a screen, allowing others to project their own image of what happened in this distant place.

And yet nobody denies that many miners died from small-cell lung cancer. The epidemic was first documented in 1956, when government health officials had an autopsy performed on Tom Van Arsdale, a fifty-one-year-old miner. Experts recommended that mines ban smoking, improve ventilation, and institute other safety features. But a culture of secrecy remained from the war years, and agencies buried the reports. In Colorado, stricter regulations weren't instituted for a decade, and it took even longer for victims and their families to receive substantial restitution. In the 1970s, this battle was taken up by Stewart Udall, the Secretary of the Interior under Kennedy and Johnson. Udall represented families of Navajo Indians who died after working in terrible conditions in New Mexico; in Udall's words, the government "had needlessly sacrificed the lives of [the Navajo miners] in the name of national security." In 1990, Congress finally passed the Radiation Exposure Compensation Act, which provides health care and cash payments of $150,000 and more to miners and other uranium workers who become ill.

But the cover-up doesn't seem to cause much anger in southwestern Colorado. In the town of Naturita, I met Marie Templeton, a local historian and the daughter of Tom Van Arsdale. Templeton's husband also died of small-cell lung cancer, but she refused to see the men as victims. She told me that they had chosen a high-risk occupation that paid well, and they noticed a trend of health problems among colleagues long before the story broke. "They knew," she said. "It was an accepted risk, because they were earning a good living for their families." Like everybody I met who had lost a family member, Templeton supported the new mill, and like many locals she was a sometime hoarder of uranium. She hoped the industry would return full-bore. "All of those safety features have been built in now," she said. "Anyway, if you have a pile of high-grade uranium ore laying on the ground, you can go out there and roll in it and it wouldn't hurt you. That's a fact. You can check that. I had a nice piece of ore for forty years or better. It was in a close environment in my home, but nothing happened to me."

Even sick miners speak the same way. Billy Clark, who has pulmonary fibrosis, told me that he'd be glad to see the industry return, because now it's better regulated. But he shook his head when I asked about his old coworkers. "Most of them goddam gone," he said. "The ones that are around, they're worse than me. They're on oxygen."

His wife, Debbie, spoke up. "It gets to where your lungs crystallize," she said. "My uncle passed away last year. His lungs just crystallized and he was spitting up this bloody stuff. They told us it was parts of his lungs."

Billy used oxygen only at night. This seems to be a point of pride among former miners, who rely on their tanks as little as possible. They often note that breathing in is easy—the hard part is breathing out. This distinction appears to be important to them, as if it might cut any unwanted sympathy in half. In conversation, they sometimes drift away from the topic of health problems in order to reminisce about old mines, whose names are pronounced as fondly as lost loves. "That Golden Cycle," Billy said with a smile. "That stuff was so hot it'd cake up on the walls. When it was hot like that it was sticky." He was in his early sixties, and he had started working underground as a sixteen-year-old; people in town still called him by his mining nickname of "Hard Rock." He said the pay had always been good. I asked if he had saved much money.

"Hell no," he said. "They didn't put handles on it. I spent it mostly on good times. Aw, what the hell. You only have fun once."

These conversations were never tense. Domestic scenes followed a certain pattern: a husband who couldn't breathe, a wife who helped him remember, both of them chatting about the topic of sickness and death as calmly as if it were tomorrow's weather. People often lived in a house or a trailer that was bought with a government settlement. Pat Mann put a new roof on her home after her husband died; Billy Clark's double-wide was paid for with what his wife jokingly referred to as "blood money." People commonly expressed gratitude for Union Carbide.

"The way I look at it, I wanted a job," Larry Cooper told me. He had worked for years in Uravan; his mining nickname was Coop. "They gave me a job. I didn't ask if I was going to get cancer—which I did. It was just one of those things. I think our life wasn't too bad, don't you, Mom?"

"Not at all," said his wife, Avis. We were sitting in the living room of their house in Nucla, where they had moved after leaving Uravan. Avis was knitting an afghan with the colors of the American flag. Coop told me that half of his right lung had been removed. "But I smoked probably sixty years," he said. "So I won't say that the cancer was caused by Carbide."

Coop was in his early eighties, a big man in suspenders and Wrangler jeans. Whenever he breathed out, he hissed softly and pursed his lips, which gave him a thoughtful expression. He was strongly in favor of the new mill. "It's going to be so much different from what the old mills were," he said.

Avis said, "I think this green crap that they're putting out is for the birds."

"I think environmentalists are one of the ruinations of the nation," Coop said.

Avis said, "How many other people have had lung cancer that's never lived in Uravan?"

"I tell you one thing, I ain't been worth a damn since that surgery," Coop said. "My trouble is, I can get air in, but I can't get rid of it." He continued, "I smoked everything that was ever built. Dominoes, Avalon, Sensation, Wings. And I worked in the mines. I'd say the two of them caused it, don't you, Mom?"

"This idea that radiation causes lung cancer," Avis said. "All these millions of people who got lung cancer, they didn't live in Uravan."

"I can't say nothing bad about Carbide, and nothing bad about mining," Coop said. He pursed his lips and hissed thoughtfully. "Let's put it this way," he said. "You're going to live until you die. And there's nothing you can do about that."

IT'S NEVER A GOOD SIGN WHEN A COMMUNITY POSTS A "For Sale" notice in front of its elementary school. The town of Nucla, which is about fifteen miles from the proposed mill site, finally put its old school building on the market in 2009, because there are so few children left. The local high school used to graduate eighty students annually; last year there were eleven seniors. The surrounding region once had a population of between six and eight thousand; now there are only about sixteen hundred people. Many locals work construction or cleaning jobs in Telluride, the prosperous ski community that is about sixty-five miles from the proposed site of the Paradox mill.

Much of the opposition to the mill has come from Telluride. The Sheep Mountain Alliance, an environmentalist group, has filed a lawsuit to halt the project, but activists acknowledge that it's hard to oppose development when you come from a thriving town. "I recognize how patriarchal that can seem," Hilary White, the director of the Sheep Mountain Alliance, told me. She believed that economic problems made people receptive to the uranium industry, which she didn't think could be trusted to follow regulations. "When you're desperate, when you can't afford to put food on your table, you'll welcome people who don't have your best interests at heart."

For outsiders, the local relationship with history can seem baffling, especially with regard to Uravan. "We thought that was such an obvious way to fight it, by referring to Uravan," White said. "These people lived through it; they saw it. But this is all they know." At a public meeting in the county seat of Montrose, the actress Daryl Hannah spoke out against the mill; she owns a home between Telluride and Paradox. "It's kind of mind-boggling for me to hear people say, 'I worked at the Uravan mill, and it was a booming economy at the time, and I wish we could go back to it,'" Hannah told a reporter. "But you look at Uravan now and it's completely fenced off, and it says 'Radioactive, Do Not Enter, Dangerous, Use Caution!'"

Uravan thrived during the 1960s and 1970s, when the American uranium industry shifted from defense to energy production. But in the 1980s the town laid off workers, and after Three Mile Island there was increased public concern about radioactive sites. The State of Colorado successfully sued Union Carbide, forcing a Superfund cleanup so extensive that it required the town's destruction. All remaining residents were moved out—the last to go was the postmistress—and on New Year's Eve of 1986 Uravan was officially closed.

For the next two decades, Union Carbide, which was acquired by Dow Chemical in 2001, worked in conjunction with the federal government in trying to remove virtually all traces of radioactive contamination from the site. A crew of as many as a hundred workers demolished the mill, the school, the houses—a total of 260 structures. All local roads were torn up. After that, the cleanup proceeded into the soil, like an archaeology of the atomic age. Workers found a vial that was believed to have contained radium, once the world's most expensive substance. They discovered that one section of State Highway 141 passed directly over the site of the Manhattan Project's uranium mill, so they tore up the new road and replaced it. They excavated the foundation of the first radium mill, the one intended to cure cancer, and they destroyed that, too.

Regulations specified that everything had to be shredded and buried in four repositories atop a neighboring mesa. Uravan contained some working bulldozers, dump trucks, and Caterpillar loaders; all were ripped apart with hydraulic shears. Storerooms of unopened supplies—sinks, toilets, test tubes, lightbulbs, whatever—were torn to bits. One worker told me he shredded a brand-new stainless-steel rod worth at least five thousand dollars. They ripped up water lines; they uprooted gardens; they tore down every tree in town. Whenever equipment left the site, it was washed and checked with a Geiger counter. If a bulldozer blade or a hydraulic shear couldn't be cleaned to strictly low radiation levels, it was destroyed on the spot. Sometimes tires had to be removed and shredded.

To design the repositories, computers simulated the worst storm that was likely to occur in the next thousand years. The Uravan site is now fenced in, with warning signs that read, "Any Area or Container on This Property May Contain Radioactive Materials." Soon, it will become the property of the Department of Energy, which intends to keep it closed for all eternity.

The destruction of Uravan took more than two decades, and it cost $127 million, of which about $50 million came from federal funds. On the other side of the world, Hiroshima and Nagasaki are thriving cities, but the town that helped make the bomb has been wiped completely off the face of the earth.

TODAY, NOBODY LIVES WITHIN NINE MILES OF THE SITE, WHICH is a narrow plain pressed between two cliffs of sandstone. Each summer, former Uravan residents hold a reunion picnic nearby, and occasionally people stop by on their own, as if it were a cemetery. Whenever I accompanied visitors, there was a strange nostalgia: people leaned over the fence, which was covered with radioactive warning signs, and pointed out where they had been married, and where children had been born, and where teenage indiscretions had been committed. "I kissed a boy on the bridge on Halloween!" a woman in her fifties said, giggling. "That was the original tailings pile," a man said fondly, pointing at an empty space beside a cliff. "We'd take old car hoods and slide down it."

In a region of tough people, this obscure and ruined spot sometimes seemed the sole point of sentimentality. Twice, former residents wept while talking about Uravan, which never happened during a discussion of a family death from cancer. Some of this had to do with the perception of agency: locals found dignity in mining and milling, which reflected personal decisions, whereas nobody had had a choice in leaving Uravan. But there was also a strong sense of injustice and waste. People loved the town, and they believed the cleanup was unnecessary; they hated the way outsiders assumed that Uravan natives suffered birth defects and other health problems. "$U_3O_8$, when it's first mined, it's not that hot," Gene Greenwood, a former Uravan resident who had helped supervise the cleanup, told me, using the chemical abbreviation for yellowcake. He noted that people tend to conflate different forms of uranium—yellowcake, enriched fuel, and bomb material—when in fact each has a markedly different production process and radioactivity level. He said that the Uravan cleanup hadn't been motivated by health concerns, and now the site is no more radioactive than the surrounding landscape. "It was a liability issue," he said. "It wasn't a safety issue."

The locals often speak of uranium and nuclear power in highly technical terms. They refer to "thermoluminescent dosimeters," and they distinguish between alpha and gamma radiation. The phrase "pulmonary fibrosis" often crops up. The sudden sophistication can be jarring; once, a woman compared President Obama to Adolf Hitler and then mentioned the results of an epidemiological study on the effects of ionizing radiation. Few people have much formal education, and they have a reputation for being insular; in this remote place, folks often adopt a worldview that is fearful of the outside. But anything atomic seems to make them comfortable. They aren't scared of radiation, and they often say surprising things. Several people insisted to me that unenriched uranium isn't carcinogenic, and they said there's no evidence that the low radiation levels involved in regulated mining and milling have negative health effects. Howard Stephens, who had worked in the Uravan mill, told me that the radiation levels he received there were about the same as those of someone employed in New York's Grand Central Terminal. People said that the airline industry exposes employees to more radiation than the nuclear industry does. Ron Henderson, a county commissioner, told me that yellowcake is so harmless that it can be mailed in the U.S. postal system. "It's like sending powdered sugar," he said. "You put it in a Ziploc. You just want to make sure you zip it all the way across."

The contrast with the environmentalists who opposed the mill couldn't have been greater. They were better educated and more worldly, and their opinions weren't influenced by the prospect of financial gain. But I noticed a vagueness with regard to scientific issues. "There's always been a lot of talk of leukemia and cancer rates around these places," Travis Stills, a lawyer who is involved with two of the anti-mill lawsuits, told me. When I asked about evidence, he said that epidemiological studies were unreliable. Activists often quoted a statement issued by the Larimer County Medical Society in Fort Collins, Colorado, which claimed that communities engaged in uranium mining have suffered a "documented increase" in leukemia, childhood bone cancer, miscarriages, genetic abnormalities, and other serious conditions. But when I contacted the physicians who issued the statement they couldn't produce a source. (One told me that he had lost his materials to water damage.)

And yet almost everything I heard in the uranium towns could be documented. The World Health Organization does not classify uranium as a human carcinogen. The walls of Grand Central Terminal are made of granite, which contains elements that produce radon; a worker there receives a larger dose of radiation than the Nuclear Regulatory Commission allows a uranium mill to emit to a next-door neighbor. Being closer to the sun—living in the mountains, flying in planes—also means more radiation. According to the National Council on Radiation Protection, the average airline crew member receives an annual dose of work-related radiation that is more than one and a half times higher than that of the average employee in the nuclear power industry. (Neither dose is higher than what the typical American receives from natural background radiation.) And there is no compelling evidence that low amounts of radiation cause health problems. Finally, I wondered if even the craziest things I'd heard were true, and I called the regional spokesman for the U.S. Postal Service. He told me in no uncertain terms that yellowcake is classified as a UN2912 radioactive material, and that it is strictly forbidden in the mail, regardless of whether you zip the bag all the way across.

URAVAN HAS BEEN STUDIED BY DR. JOHN BOICE, WHO founded the radiation epidemiology branch of the National

Cancer Institute. He now teaches at Vanderbilt School of Medicine, and is also the scientific director of the International Epidemiological Institute, an independent research organization. When I met him at his office in Rockville, Maryland, he told me that Uravan, like many company towns, kept unusually good records. This allowed Boice and other researchers to find out who lived in the town and where they worked. Accessing data from 1936 to 2004, they traced mortality for a wide range of diseases. "We did find a significant increase of lung cancer," he said. "But in men only. It was concentrated in the miners. There was nothing with the women who lived in town. If you're looking at an environmental exposure, you'd expect men and women to be the same. And even the millers did not show this."

The overall Uravan mortality rate was 10 percent lower than the national average. There was less heart disease, which probably reflected the lifestyle of well-employed people who liked outdoor activities. Boice had conducted other studies in uranium regions, and the only significant risks he had found involved miners who labored in unventilated conditions, especially those who smoked. He noted that safety measures have made an enormous difference. "The radon levels are so low now," he said. "In the early days, they had no standards."

Activists told me that Boice's Uravan study was unreliable, because it had received funding from Union Carbide. When I had the material reviewed by independent experts, they said the methodology was sound, and the findings were in line with those of the National Institute for Occupational Safety and Health, which has found no statistically significant evidence that mill workers suffer higher mortality rates because of either the radiation or the chemical toxicity of uranium. In New Mexico, where the DiNEH Project is studying Navajo communities, some with a history of unregulated mining, researchers told me that they believe there is an association with kidney disease and other ailments, but they cautioned that the findings are still preliminary.

Scientists said that, despite the public perception, radiation is a weak carcinogen. In the 1980s, the National Cancer Institute conducted an extensive study of all 107 American counties that contained a nuclear power plant or a Department of Energy nuclear facility. The study found no excess cancers. Recently, the institute participated in a fifteen-nation study that involved over four hundred thousand nuclear industry employees, all of whom had worn dosimeters that tracked radiation levels over years of work. Dr. Ethel S. Gilbert, a scientist at the institute, told me that they found no evidence of increased mortality for people exposed to doses of less than 0.1 sieverts, which is more than fifty times the average annual dose of an American nuclear-power employee. She talked about the challenges of explaining such issues, because people fail to distinguish between high and low doses of radiation. "They think if you get exposed it's bad," she said. "It's hard to understand that the dose is important." Gilbert described what researchers know about an exposure of 0.1 sieverts, which was found among only 5 percent of study subjects, mostly from the early years of the nuclear power industry. From the industry perspective, such a dose is high, but not in terms of health effects. "Out of one hundred people exposed to 0.1 sieverts, we would expect one cancer from that exposure," Gilbert said. "But there would also be forty-two people who would get cancer for other reasons. It's very hard to study these low levels, because there are so many other things that contribute to people getting cancer."

The effects of high doses are well documented, largely because of a sixty-year study of nearly a hundred thousand Japanese atomic-bomb survivors. With high levels of radiation, there's a clear linear pattern—more exposure means an incremental rise in risk. But it's unclear whether this pattern continues into the lower-dose range, where any health effects are so small that they can't be demonstrated by epidemiological studies. Some experts and scientific bodies, including the French

Academy of Sciences, have questioned the linear model for low levels, believing that radiation may be harmless up to a certain threshold. This is common for many elements and environmental factors—iron and zinc, for example, are healthy up to a certain threshold, but in high doses they are poisonous. Applying such a model to radiation, though, is controversial, because it would radically change risk assessment, as well as possible solutions for the storage of nuclear waste.

United States regulations continue to follow the linear no-threshold theory. It has the benefit of being simple and safe, but it can also be misinterpreted. Because of Colorado's elevation, a resident there receives two to three times the natural background radiation of someone who lives in New Jersey, so strictly speaking there should be an increased risk of cancer. (In fact, Colorado cancer rates are lower.) After the Chernobyl accident, in 1986, anti-nuclear groups and scientists used the findings from the Japanese atomic-bomb survivors, extrapolated downward for the radiation levels in Europe, and predicted tens of thousands of deaths from cancer. Critics note that this is like taking a set of deaths from motorists who drove a curve at a hundred miles an hour and making the assumption that, if people slow to ten miles an hour, they'll die at a tenth of the original rate. This is also why $127 million was spent obsessively cleaning up an abandoned town whose former residents lived longer than the national average. Metaphorically speaking, the Uravan speed limit was set at one.

Even worst-case disasters reveal surprisingly small effects. In Chernobyl, dozens of emergency workers died after fighting the reactor fire, but the health impact on neighboring communities seems to be limited. After more than twenty years of extensive study, there is no consistent evidence of increased birth defects, leukemia, or most other radiation-related diseases. The only public epidemic consists of high rates of thyroid cancer in children, whose glands are particularly sensitive to radiation. Fewer

than ten people have died—thyroid cancer is usually treatable—
but many have had to undergo surgeries, and it will be years
before the full impact of the epidemic is known.

Even this epidemic, like the accident itself, could have been
avoided entirely. The Soviet reactor lacked a containment facility, a design flaw that is unimaginable today, and the Communist government delayed announcing the accident. "The Russians could have done one thing that would have gotten rid of
the epidemic of thyroid cancer," Boice told me. "They could have
said, 'Don't drink the milk.'" In surrounding areas, cows ate
grass contaminated by fallout, and people fed the milk to their
children. An open society would have responded differently;
even as far back as 1957, when a fire at a badly designed British nuclear facility called Windscale released radiation, all local
milk was dumped into the sea. In 2011, when an earthquake
and tsunami caused a partial meltdown in two nuclear reactors
in Japan, there was no public epidemic, because residents were
evacuated and food was monitored for contamination. Despite
the fact that the Japanese reactors had been poorly managed
and maintained, they released only a sixth of the radiation of
Chernobyl—the containment facilities helped prevent a disaster. There's been no evidence that the Japanese public was exposed to dangerous levels of radiation, and of the four thousand
people who worked on the plants in the wake of the disaster,
only one hundred and three were found to have received an exposure of more than 0.1 sieverts. At that level, scientists would
predict a 1 percent increase over the normal cancer rate. It's a
relatively small effect, especially in the context of a tsunami that
killed more than twenty thousand—but the nuclear meltdown is
all that most people remember.

Boice told me that the biggest health problems from high-
profile accidents are often psychological. A twenty-year study
showed no consistent evidence that the low amounts of radioactivity released in the Three Mile Island accident have

had a significant impact on mortality in communities around the reactor. But people suffered from high rates of stress and increased alcohol consumption. Places near Chernobyl have high rates of alcoholism, tobacco use, and depression. After the Ukrainian accident, European countries as distant as Greece reported a significant spike in elective abortions, owing to a fear of birth defects. Because of Chernobyl, a number of European nations scaled back dramatically on nuclear power, and Italy closed down all its reactors. Twenty years later, Italy purchases electricity from France, which is 80 percent nuclear, and which ranks twenty-fourth out of twenty-seven European Union countries in terms of absolute greenhouse gas emissions.

None of the cancer specialists I spoke with opposed nuclear power for health reasons. Invariably, their biggest worry was the storage of nuclear waste, although many people note that this problem is more political than scientific. Several scientists told me that there should be greater public concern about medical radiation, because high-dose procedures like CT scans can be overprescribed, and regulation is light in comparison with the nuclear power industry. (From 1996 to 2006, the number of CT scans performed in the United States increased nearly threefold.)

Boice expressed concerns about terrorism, but largely because he believes people are seriously misinformed about radiation. Converting yellowcake or even enriched reactor fuel into an effective bomb is complex and probably impossible for a terrorist group, but that's not the issue. Even materials with low levels of radioactivity—for example, the kind of stuff you find in gardens and living rooms in southwestern Colorado—would terrify most people. "We've studied radiation for one hundred years," Boice told me. "We know a lot about it. But it's invisible. A colleague said, 'If you could paint it blue and see it, it wouldn't be such an issue.'"

*   *   *

WHILE MOST GRASSROOTS ENVIRONMENTALISTS REMAIN ANTI-nuclear, the evidence of climate change has led some prominent greens to become vocal supporters. Patrick Moore, one of the founders of Greenpeace, left the organization, and believes that it takes an unscientific view of the issue. The Gaia theorist James Lovelock is a particularly outspoken advocate, as is Stewart Brand, the founder of the *Whole Earth Catalog*. In 2009, Mark Udall, the Democratic senator from Colorado, addressed the issue on the Senate floor. "For some," he began, "news that a Udall is speaking favorably about nuclear power will come as a stark—and perhaps unpleasant—surprise." Udall comes from a prominent family of environmentalists, and his uncle, Stewart Udall, who died in 2010, represented the Navajo uranium miners in their lawsuit. Senator Udall told me that he deeply respects his uncle's legacy, but that current regulations are much improved and the threat of climate change requires new thinking. "The risks that are attendant to the use of nuclear power are worth embracing," he said. "Just like the risks that come with increasing the use of natural gas, or frankly even renewables."

Udall prefers wind and solar energy, but he acknowledged that they aren't capable of significantly displacing coal in the near future. He believes there are solutions to the issue of nuclear-waste storage. When I mentioned the proposed mill in his home state, he said, "I can support such a project if all the requisite laws and regulations are met." He continued, "My uncle wrote very compellingly about the nature of the Cold War, and how we were driven into a mindset that we have to be secretive here, we don't need any regulation or controls, because that would somehow threaten the very existence of America. We've moved into another era."

I asked whether fear of climate change, like the old fear of the Soviets, might lead to rash decisions and carelessness. "Given the difficulty that we face right now in Washington in convincing a supermajority in the Senate that we have to respond to climate change," Udall said, "I'm not as worried about that. But I think it behooves us to never forget these lessons."

THERE'S A TIMELESS QUALITY TO THE FAR CORNER OF COLOrado, which is too remote for national chains, and where travelers stay at the Ray Motel in Naturita. The Ray still uses keys, and the key chain still has a 1970s-era message promising that if it's dropped into any U.S. mailbox it will be returned free of charge. In January, when I checked in, a receptionist named Sherri Ross asked if I had arrived for the public hearing on the Energy Fuels mill. Ross explained that she was a former Uravan resident whose father and several uncles had all died from mining-related lung cancer. By now, this information was sufficient for me to guess that she was a wholehearted supporter of the industry. "We've had the biggest loss you can ever have, and we're not against the mill," she said. A cleaning lady walked past and commented that lung cancer had also killed her father, and she sure hoped the mining would come back.

The Colorado Department of Public Health and Environment was holding the hearing. Most people predicted that the health department would eventually grant approval, but lawsuits were still pending and others would likely be filed. The real issue, though, seemed to be financial. Energy Fuels is listed on the Toronto Stock Exchange, and the share price had plummeted in the past year. There have been some signs of a nuclear resurgence, but the United States' path remains unclear. Demand may be more likely to come from overseas, especially India and China, which have announced ambitious plans for nuclear power. At the moment, China relies heavily on power from coal-burning plants and hydroelectric dams, and the vast majority of the population lives in the south and the east, where the possibilities for wind and solar energy are not promising.

George Glasier, the founder and CEO of Energy Fuels, told me that he was confident there would be demand. He had worked as a lawyer for uranium companies until the crash in the 1980s, when he bought a ranch in southwestern Colorado. Like many industry executives, he was savvy about riding the economic waves. A number of years ago, he had opened a gravel pit on his property, selling rock for use in covering the remains of Uravan. He had also sold a hundred thousand cubic yards of topsoil that was poured onto another contaminated mill site. Now that the industry seemed ready to shift from obsessive cleanup to real production, Glasier hoped to get back in business. He still kept a chunk of ore and a jar of yellowcake in his home. "That's fairly high grade," he said, handing me the rock. I didn't open the jar.

For branding purposes, Energy Fuels had named its project the Piñon Ridge Mill. "We didn't want to have the word 'paradox' in there," Glasier said. He produced a promotional display from the Nuclear Energy Institute: a plastic pellet that, if it were real enriched uranium, would have the generating capacity of a ton of coal. The display said: "Nuclear. Clean Air Energy." It dated to the 1970s, when people worried about smog instead of climate change; the font was as old-school as the Ray Motel's key chain. "I'd give it to you, but I don't have any more left," Glasier said.

The hearing was held in the nearby town of Nucla, which is about fifteen miles from the mill site. Such meetings have been scheduled throughout the region, where reactions follow a pattern: the farther you go from the mill, the more frightened people seem. And there's no telling what will be said when an open mic is offered in southwestern Colorado. At one meeting in the county seat, a man accused the health department of trying to murder citizens. Another time, a speaker reminisced about how great the tomatoes tasted in radioactive Uravan. He was followed by a man who announced, apropos of nothing, "I'm not really into a black guy for president." Environmentalists said they often felt uncomfortable at such meetings, where locals sometimes express anger toward opponents. I sympathized with the outsider perspective—as a writer, I often inspired the same response. But I came to understand the reason for this anger. Natives were accustomed to condescension, especially with regard to health issues, when in fact their specialized knowledge ran deep. I took many local opinions with a grain of salt, but I learned to listen when people talked about uranium.

At the Nucla meeting, more than two hundred citizens attended, and the vast majority wore orange buttons that said "Yes to Mill." "We're not afraid of uranium here," Joyce Shaffer, a former Uravan resident, said into the microphone. "I don't like skiing. I'm afraid of it. I don't understand it. But I understand uranium, and I'm not afraid of it." Another woman identified herself as a fourth-generation resident. "I've had family members who passed from working in the nuclear industry," she said. "I have no regrets for that." A member of the Chamber of Commerce made a statement that could only come from a region called Paradox: "Uranium and tourism can coexist."

It wasn't until the thirtieth speaker that somebody opposed the project. In the end, only five spoke out against it: they referred to health risks, wildlife issues, and the storage of nuclear waste. Craig Pirazzi, a Paradox resident, criticized the industry's volatility. "These are not stable jobs," he said. "These people deserve better than this."

EVERY AUGUST, URAVAN NATIVES RETURN FOR THEIR PICNIC. With the site fenced off, they congregate on the former baseball diamond, which is a mile to the southeast. The dugouts are long gone, and tufts of switchgrass have overrun the base paths, but there was never a cleanup here—no wire fence, no warning signs. The site is pleasantly shaded by big cottonwoods that escaped destruction.

At the 2010 picnic, George Glasier told me he was happy to be ranching full-time. A few months earlier, he had stepped down as the CEO of Energy Fuels, announcing that the company needed new direction. Initially, the stock price dropped to twelve cents before rebounding a bit. Most people believe that the company hopes to get licensing approval and then sell out to a big corporation that can weather the uncertainty. It's another timeless quality in the uranium towns: they might be thirty years too late, but it's also possible that they're ten years too early.

More than two hundred people came to the picnic. They traveled from as far away as Houston and Los Angeles, and a couple of Navajo families drove up from New Mexico. Two former Uravan doctors showed up, and the high-school class of 1969 held its fortieth reunion. Many people wore T-shirts that read "DANGER: Radioactive Material: I Lived in Uravan, Colorado!" An organization called Professional Case Management distributed information about government assistance for former uranium workers.

At the base of one cottonwood, people arranged old metal street signs from Uravan: Flint Avenue, Mill Drive, Calcite Avenue. They were supposed to have been shredded along with everything else, but workers had sneaked them out. A man named Stan Cadman, who had grown up in town, brought an enormous "Uravan" sign that once marked the highway turnoff. Cadman now drives a trucking route through southwestern Colorado, and often, in the middle of the night, he pulls over where his hometown once stood.

"It's haunted up there," he told me. "You can hear voices." I thought about everything that might haunt such a place—mill sounds, coughing old miners, maybe even Japanese targets of the bomb—and I asked Cadman what he heard. He was a big man with a Harley-Davidson cap, tattooed forearms, and a biker beard. He smiled and said, "You can hear kids playing."

# STRANGE STONES

ALL ALONG HIGHWAY 110 WE SAW SIGNS FOR STRANGE STONES. They first appeared in Hebei Province, where the landscape was desolate and the only color came from the advertising banners posted beside the road. They were red and had big characters promising *qi shi*—literally, "strange stones," although the adjective could also be translated as "marvelous" or "rare." The banners had been tattered and torn by the wind. We were driving northwest, right into a spring storm. There was only rain at the moment, but we could see what lay ahead—the forecast was frozen on top of the oncoming traffic. Most vehicles were big Liberation-brand trucks carrying freight south from Inner Mongolia, and their stacks of boxes and crates were covered with ice. The trucks had fought a crosswind on the steppes and now their frozen loads listed to their right, like ships on a rough sea.

I was driving a rented Jeep Cherokee, and Mike Goettig was along for the ride. If things went well, I might eventually make it to the Tibetan Plateau. We had met in the Peace Corps, and after finishing our time as volunteers we had each found a different way of staying in China: I worked as a freelance writer; Goettig opened a bar in the southwest. But every once in a while we met up on the road, for old times' sake. We passed a half-dozen signs for Strange Stones before either of us spoke.

"What's up with this?" Goettig said at last.

"I have no idea. I haven't driven this road before."

The banners stood in front of small shops made of cement and white tile, and they seemed to grow more insistent with every mile. "Strange stones" is the Chinese term for any rock whose shape looks like something else. It's an obsession at scenic destinations across the country; in the Yellow Mountains you can seek out natural formations with names like Immortal Playing Chess and Rhinoceros Watching Moon. Collectors buy smaller rocks; sometimes they've been carved into a certain shape, or they may contain a mineral pattern with an uncannily familiar form. I didn't have the slightest interest in Strange Stones, but their proliferation in this forgotten corner of Hebei mystified me. Who was buying this stuff? Finally, after about twenty banners, I pulled over.

Inside the shop, the arrangement seemed odd. Display tables completely encircled the room, leaving only a narrow gap for entry. A shopkeeper stood beside the gap, smiling. With Goettig behind me, I squeezed past the tables, and then I heard a tremendous crash.

I spun around. Goettig stood frozen; shards of green lay strewn across the concrete floor. "What happened?" I asked.

"He knocked it off!" the shopkeeper said. He grabbed the hem of Goettig's coat. "Your jacket brushed it."

Goettig and I stared at the scattered shards. Finally I asked, "What is it?"

"It's jade," the man said. "It's a jade ship."

Now I recognized pieces: a corner of a smashed sail, a strand of broken rigging. It was the kind of model ship that Chinese businessmen display in their offices for good luck. The material looked like the cheap artificial jade that comes out of factories, and the ship had exploded—there were more than fifty pieces.

"Don't worry about it," the shopkeeper said brightly. "Go ahead and look around. Maybe there's something else you'll want to buy."

We stood in the center of the room, surrounded by the ring of tables, like animals in a pen. Goettig's hands were shaking; I could feel the blood pulsing in my temples. "Did you really knock it over?" I said, in English.

"I don't know," he said. "I didn't feel anything, but I'm not sure. It fell down behind me."

I had never seen a Chinese entrepreneur react so calmly when goods were broken. A second man emerged from a side room, carrying a broom. He swept the shipwreck into a neat pile, but he left it there on the floor. Silently, other men appeared, until three more of them stood near the door. I was almost certain it was a setup; I had heard about antique shops where owners broke a vase and blamed a customer. But we were hours from Beijing and I didn't even know the name of this county. Goettig had become extremely quiet—he was always like that when something went wrong. Neither of us could think of a better plan, so we started shopping for Strange Stones.

GOETTIG AND I HAD BOTH JOINED THE PEACE CORPS IN 1996, when it seemed slightly anachronistic to become a volunteer. The organization's character has always shifted with the American political climate, ever since President John F. Kennedy founded it in 1961, during the heart of the Cold War. Back then, the Peace Corps became immensely popular, attracting idealistic young people who were concerned about America's role in the developing world. Later, after the Vietnam War, the organization suffered as the nation experienced a wave of cynicism about foreign policy. Since the attacks of September 11, the significance of the Peace Corps has changed again—nowadays anybody who joins is likely to have thought hard about personal responsibility in a time of war.

During the mid-1990s, though, there were no major national events that weighed on volunteers. It was hard to say what motivated a person to spend two years abroad, and we came for countless reasons. Most of the volunteers I knew possessed some strain of idealism, but usually it was understated, and often people felt slightly uncomfortable speaking in such terms. Goettig told me that during his interview with the Peace Corps the recruiter had asked him to rate his "commitment to community" on a scale of one to five. Goettig gave himself a three. After a long pause, the recruiter started asking questions. You've worked in a drug-treatment center, right? You're teaching now, aren't you? Finally he said, "OK, I'll put you down as a four." Goettig told me later that one reason he signed up was that he had a girlfriend in Minnesota who wanted to get serious. I heard the same thing from a few other volunteers—the toughest job you'll ever love was also the easiest way to end a relationship.

Back then, I wouldn't have told a recruiter my own true motivations. I wanted time to write, but I didn't want to go to school anymore and I couldn't imagine working a regular job. I liked the idea of learning a foreign language; I was interested in teaching for a couple of years. I sensed that life in the Peace Corps would be unstructured, which appealed to me; but they called it volunteerism, which would make my parents happy. My mother and father, in Missouri, were Catholics who remembered Kennedy fondly—later I learned that the Peace Corps has always drawn a high number of Catholics. For some reason, it's particularly popular in the Midwest. Of the thirteen volunteers in my Peace Corps group, six came from Midwestern states, and three were Minnesotans. It had to do with solid middle-country liberalism, but there was also an element of escape. Some of my peers had never left the country before, and one volunteer from Mississippi had never traveled in an airplane.

None of us were remotely prepared for China. Nobody had lived there or studied the language beyond a few basics; we knew virtually nothing about Chinese history. One of the first things we learned was that the Communist Party was suspicious of our presence. We were told that during the Cultural Revolution, the government had accused the Peace Corps of links with the CIA. These things were no longer said publicly, but some factions in the Chinese government were still wary of accepting American volunteers. It wasn't until 1993 that the first Peace Corps teachers finally showed up, and I was part of the third group.

We must have been monitored closely. I've often wondered what the Chinese security officials thought—if our cluelessness confused them or simply made them more suspicious. They must have struggled to figure out what these individuals had in common, and why the United States government had chosen to send them to China. There were a few wild cards guaranteed to throw off any assessment. A year ahead of me, an older man had joined up after retiring from the U.S. Coast Guard. Everybody called him the Captain, and he was a devoted fan of Rush Limbaugh; at training sessions he wore a Ronald Reagan T-shirt, which stood out on the Chinese college campus where he lived. At one point, a Peace Corps official said, "Maybe you should change your shirt." The Captain replied, "Maybe you should reread your Constitution." (This was in the city of Chengdu.) One day, while teaching a class of young Chinese, the Captain drew a line down the heart of the blackboard and wrote "Adam Smith" on one side and "Karl Marx" on the other. "OK, class, short lesson today," he announced. "This works; this doesn't." In the end, the Peace Corps expelled him for breaking a cabby's side-view mirror during an argument on a Chengdu street. (This altercation happened to occur on Martin Luther King Jr. Day, a nice detail that probably escaped the Chinese security file.)

After a while, it was almost possible to forget who had sent you and why you had come. Most of us taught at small colleges in remote cities, and there wasn't much direct contact with the Peace Corps. Only occasionally did a curriculum request filter down from the top, like the campaign for Green English. This was a worldwide project: the Peace Corps wanted educational volunteers to incorporate environmental themes into their teaching. One of my peers in China started modestly, with a debate about whether littering was bad or good. This split the class right down the middle. A number of students argued passionately that lots of Chinese people were employed in picking up garbage, and if there wasn't any litter they would lose their jobs. How would people eat when all the trash was gone? The debate had no clear resolution, other than effectively ending Green English.

The experience changed you, but not necessarily in the way you'd expect. It was a bad job for hard-core idealists, most of whom ended up frustrated and unhappy. Pragmatists survived, and the smart ones set small daily goals: learning a new Chinese phrase or teaching a poem to a class of eager students. Long-term plans tended to be abandoned. Flexibility was important, and so was a sense of humor. There had been nothing funny about the Peace Corps brochures, and the typical American view of the developing world was deadly serious—there were countries to be saved and countries to be feared. That was true of the Communists, too; their propaganda didn't have an ounce of humor. But the Chinese people themselves could be surprisingly lighthearted. They laughed at many things, including me: my nose, the way I dressed, the way I spoke their language. It was a terrible place for somebody stiffly proud to be American. Sometimes I thought of the Peace Corps as a reverse refugee organization, displacing all of us lost Midwesterners, and it was probably the only government entity that taught Americans to abandon key national characteristics. Pride, ambition, impatience, the instinct to control, the desire to accumulate, the missionary impulse—all of it slipped away.

AT THE SHOP, A FEW STRANGE STONES LOOKED LIKE FOOD. This has always been a popular Chinese artistic motif, and I recognized old favorites: a rock-hard head of cabbage, a stony strip of bacon. Other stones had been polished to reveal some miraculous mineral pattern, but in my nervousness they all looked the same to me. I selected one at random and asked the price.

"Two thousand yuan," the shopkeeper said. He saw me recoil—that was nearly $250. "But we can go cheaper," he added quickly.

"You know," Goettig said to me, "nothing else in here would break if it fell."

He was right—it was all Strange in a strictly solid sense. Why had a jade ship been there in the first place? As a last resort, I hoped that Goettig's size might discourage violence. He was six feet one and well-built, with close-cropped hair and a sharp Germanic nose that the Chinese found striking. But I had never known anybody gentler, and we shuffled meekly toward the door. The men were still standing there. "I'm sorry," I said. "I don't think we want to buy anything."

The shopkeeper pointed at the pile of green shards. "*Zenme-ban?*" he said softly. "What are you going to do about this?"

Goettig and I conferred, and we decided to start at fifty yuan. He took the bill out of his wallet—the equivalent of six dollars. The shopkeeper accepted it without a word. All the way across the parking lot I expected to feel a hand on my shoulder. I started the Cherokee, spun the tires, and veered back onto Highway 110. I was still shaking when we reached the city of Zhangjiakou. We pulled over at a truck stop for lunch; I guzzled tea to calm my nerves. The waitress became excited when she learned we were Americans.

"Our boss has been to America!" she said. "I'll go get her!"

The boss was middle-aged, with dyed hair the color of shoe black. She came to our table and presented a business card with a flourish. One side of the card was in Chinese, the other in English:

UNITED SOURCES OF AMERICA, INC.
JIN FANG LIU
DEPUTY DIRECTOR OF OPERATIONS
CHINA

Embossed in gold was a knockoff of the Presidential Seal of the United States. It looked a lot like the original, except for the eagle: the Zhangjiakou breed was significantly fatter than the American original. He had pudgy wings, a thick neck, and legs like drumsticks. Even if it dropped the shield and arrows, I doubted this bird would be capable of flight. The corner of the card said, in small print:

PRESIDENT GERALD R. FORD
HONORARY CHAIRMAN

"What kind of company is this?" I asked.

"We're in the restaurant business here in Zhangjiakou," the woman said. She told me her daughter lived in Roanoke, Virginia, where she ran another restaurant.

I pointed at the corner of the card. "Do you know who that is?"

"Fu Te," Ms. Jin said proudly, using the Chinese version of Ford's name. "He used to be President of the United States!"

"What does he have to do with this restaurant?"

"It's just an honorary position," Ms. Jin said. She waved her hand in a way that suggested, *No need to tell Mr. Fu Te about our little truck stop in Zhangjiakou!* She gave us a discount and told us to come back any time.

We stopped in the city of Jining for the night. The temperature had plummeted into the teens; the rain had turned to snow; I pulled into the first hotel I could find. It had a Mongol name—the Ulanqab—and the lobby was so big that it contained a bowling alley. We registered at the front desk, surrounded by the crash of balls and pins, and by now I had a pretty good idea where this trip was headed.

TRAVELING WITH GOETTIG WAS A CALCULATED RISK. INTER-esting things happened when he was around, and he was unflappable, but his standards of comfort and safety were so low that he essentially had no judgment. Of all the Midwestern refugees I had known in the Peace Corps, he had come the farthest, and he seemed the least likely ever to return home. When our group first met for departure from San Francisco, Goettig had shown up with the smallest pile of luggage. He carried less than a hundred dollars, his entire life savings.

He was from southwestern Minnesota, where he had been raised by a single mother. She had two children by the age of nineteen, and after that she found jobs wherever she could—bartending, office work, waitressing at the Holiday Inn. Eventually, she took a position on the production line of a factory that manufactured bread-bag ties in Worthington, Minnesota, a town of ten thousand people. The family stayed in a succession of trailer courts and rental apartments; one year they lived on a farm because the previous tenant, a friend of Goettig's mother, had died in a motorcycle accident. Much of their home life revolved around motorcycles. Goettig's mother was a devoted biker, and in the summer they attended Harley-Davidson rallies and rodeos around the Midwest. He watched his mother's friends compete in events like Monkey in the Tree, in which a woman leaps from the back of a motorcycle to a low-hanging rope, where she dangles while the man continues around an obstacle course, returning so that the woman can drop down perfectly onto the street. Another contest involves seeing which woman on the back of a moving motorcycle can take the biggest bite out of a hot dog hanging from a string. When Goettig first told me about these events, I realized that I hadn't seen anything stranger in China. He said he had always hated motorcycles.

He was the only one in his family who enjoyed reading. He finished high school in the eleventh grade, because Minnesota had a program in which the state paid for a year of college if a student left the secondary system early. At the University of Minnesota at Morris, Goettig majored in English, and then he went to graduate school at the Mankato campus of the state university. While studying for his master's, he applied to the Peace Corps. He'd seen commercials as a child, and he figured it was the best way to go overseas for free.

In China, he was assigned to a job teaching English in Le-shan, a small city in southern Sichuan Province. With two other volunteers, he organized a play on the side: a student version of *Snow White*. Soon, college administrators recognized an opportunity for publicity, and they developed a traveling variety show. The other Peace Corps volunteers quickly washed their hands of the project, but Goettig was game for anything. He went on the road with *Snow White*, traveling by bus to small towns around the province, performing at middle schools three times a day. They had to change the play for political reasons. Originally, the Woodsman was a villain, but college officials insisted that the play end with a more favorable view of the proletariat, so the Woodsman reformed and gave a self-criticism. As part of the variety show, a brass band played "The Internationale," a student sang Richard Marx's "Right Here Waiting," and Goettig went onstage with a blue guitar and sang "Take Me Home, Country Roads." He was mobbed for autographs everywhere. During the bumpy rides between towns, the *Snow White* players sang songs at the top of their lungs and gorged on raw sugar cane, spitting the pulp onto the floor of the bus. Goettig told me that those were the longest ten days of his Peace Corps service.

He learned Chinese quickly. The Peace Corps gave us two and a half months of intensive training upon arrival, and after that we could hire tutors if we wished. But the best strategy was simply to wander around, talking to people in the street. Goettig had the ideal personality for this routine: he was patient and curious and tireless. He was also, as the Chinese like to say, a very good drinker. He learned to open beer bottles with his teeth, the way rural people do in Sichuan.

One autumn, he journeyed to Xinjiang, a wild region in China's far west. He camped alone in the Tian Shan Mountains, and while hiking off trail, he clambered over some rocks and was bitten on the finger by a snake. First the finger swelled, then the hand. It took four hours to make it back to Ürümqi, the provincial capital. By then, the swelling had spread to his arm, and the pain was excruciating. He found a public phone and called the Peace Corps medical officer in Chengdu. She recognized the symptoms: it sounded like a tissue-dissolving venomous snake, and he needed to get to a hospital, fast.

He asked bystanders for directions, and a young Chinese woman offered to help. She spoke perfect English, which was unusual in such a remote place, and she was dressed in a bright-orange sleeveless sweater that hung loose from her upper body like a bell. At the time, Goettig thought that the woman seemed slightly strange, but he wasn't in a position to worry about it. She escorted him to the hospital, where doctors sliced open the bitten finger. They had some traditional Chinese medicine; Goettig figured it was a good sign that the box showed a picture of a snake. The doctors used a mortar and pestle to grind up the pills, and then they shoved the powdered tablets directly into the wound.

The swelling continued to spread. Goettig's arm turned purple at the joints, where the venom was rupturing capillaries. By evening he realized that the woman in the orange sweater was completely insane. She had brought her luggage to the hospital; she refused to leave his side; she told everybody that she was his official translator. She wouldn't answer any personal questions— Goettig still had no idea how she had learned English. Whenever he asked her name, she responded, "My name is . . . Friend." Every time she said this, it sounded creepier, until Goettig finally gave up on the questions. She spent the night in a chair at the foot of his bed. The next day, the doctors cut open the hand three times to shove more powder inside. The pain was intense, but at least Goettig was able to persuade some nurses to kick the crazy woman out. After the third day the swelling began to subside. He stayed in the hospital for a week; he was so broke that the Peace Corps medical officer had to wire money to cover the bill, which was less than $150. His hand recovered fully. He never saw the woman in the orange sweater again.

A SOLITARY BOWLER WAS HAMMERING THE PINS WHEN WE checked out of the Ulanqab Hotel. At the entrance for Highway 110, the local government had erected a sign with changeable numbers, like the scoreboard at Fenway Park:

AS OF THIS MONTH
THIS STRETCH OF ROAD
HAS HAD **65** ACCIDENTS AND **31** FATALITIES

Yesterday's storm had passed, but the temperature was still in the teens. From Jining to Hohhot the highway crossed empty steppe—low, snow-covered hills where the wind howled. We passed Liberation trucks that were stopped dead on the road; their fuel lines had frozen, probably because of water in their tanks. After fifteen miles, we crested a hill and saw a line of hundreds of vehicles stretching all the way to the horizon: trucks, sedans, Jeeps. Nobody was moving, and everybody was honking; an orchestra of horns howled into the wind. Never had I imagined that a traffic jam could occur in such a desolate place.

We parked the Cherokee and continued on foot to the gridlock, where drivers explained what had happened. It all started with a few trucks whose fuel lines had frozen. Other motorists began to pass them on the two-lane road, and occasionally they encountered a stubborn oncoming car. Drivers faced off, honking, while the line of vehicles grew behind them; eventually it became impossible to move in either direction. Some had tried to go off-road, and usually they made it fifty yards before getting stuck. Men in loafers slipped in the snow, trying to dig out cars with their bare hands. There was no sign of police. Meanwhile, truckers had crawled beneath their vehicles, where they lit road flares and held them up to the frozen fuel lines. The tableau had a certain beauty: the stark snow-covered steppes, the endless line of vehicles, the orange fires dancing beneath blue Liberation trucks.

"You should go up there and get a picture of those truckers," Goettig said.

"*You* should get a picture," I said. "I'm not getting anywhere near those guys."

At last, here on the unmarked Mongolian plains, we had crossed the shadowy line that divides Strange from Stupid. We watched the flares for a while and then took the back roads to Hohhot. The moment we arrived the Cherokee's starter failed; we push-started the thing and made it to a garage. The mechanic chain-smoked State Guest cigarettes the whole time he worked on the engine, but after Highway 110 it seemed as harmless as a sparkler on the Fourth of July.

THEY SAID THE HARDEST THING ABOUT THE PEACE CORPS WAS going home. Near the end of our two years, the organization held a pre-departure conference. They handed out job-search materials, and they talked about how we might feel when we got back to America and people said things like "I didn't know they still have the Peace Corps!" A few volunteers took the foreign-service exam. One of them got halfway through and couldn't take it seriously; for the essay section he wrote about how his worldview had been influenced by the film *Air Force One*. The others passed the exam but failed the interview. Over the years, I came to know more volunteers who also took the exam, and they tended to be befuddled by the process—virtually nothing they had learned in the field seemed relevant.

From the beginning, the Peace Corps had been described as a type of foreign aid, but another goal had been to produce Americans with knowledge about the outside world. It was intended to influence national policy—the organization had been inspired in part by the 1958 book *The Ugly American*, which criticized a top-down approach to foreign affairs. At some level, I came away with a deep faith in the transformative power of the Peace Corps; everybody I knew had been changed forever by the experience. But these changes were of the sort that generally made people less likely to work for the government. Volunteers tended to be individualists to begin with, and few were ambitious in the traditional sense. Once abroad, they learned to live with a degree of chaos, which made it hard to believe in the possibility of sweeping change. The vast majority of former volunteers would have opposed the American adventure in Iraq, because their own experiences had taught them how many things can go wrong with even the simplest job. But their opinions had virtually no impact on national policy, because they didn't tend to be in positions of influence.

Many of my peers in China eventually became teachers. It was partly because we had been educational volunteers, but it also had to do with the skills we developed—the flexibility, the sense of humor, the willingness to handle anything a student could throw at you. A few became writers and journalists; some went to graduate school. Others continued to wander, and Goettig stayed in China for years. During summer, he worked for the Peace Corps, training new volunteers, and the rest of the time he picked up odd jobs: writing freelance newspaper stories, working part-time as a translator and researcher. Periodically he came through Beijing and slept on my couch for a week. The term of Peace Corps service is lifetime when it comes to guests. Sometimes I had three or four staying in my apartment, all of them big Midwesterners drinking Yanjing beer and laughing about old times.

In the southwestern city of Kunming, Goettig opened a bar with a Chinese partner. They found space in an old bomb shelter; the lease explicitly stated that they had to abandon the premises if China went to war. They had two pool tables and a stage for bands. Not long after they opened, there was a bad knife fight—one of the bartenders got stabbed multiple times, and part of a lung had to be removed. The bar didn't have much business, and Goettig and his partner barely scraped together enough money to cover the medical bills. They had named the place the Speakeasy.

The year after we drove across northern China, Goettig finally returned to the United States. He was thirty years old, and nearly broke. He went back to southwestern Minnesota, but he couldn't imagine living there again; after a month he caught a Greyhound bus heading south. Some other former volunteers were living in Starkville, Mississippi; they let Goettig crash in their home and found him a job teaching English to foreign students at Mississippi State. It paid $24,000 for the school year. When Goettig looked into teacher-certification programs, he realized that they took almost as long as law school. He bought some books about the LSAT exam, studied on his own, and scored off the charts. The next time I saw him, he was living on Riverside Drive, studying at Columbia Law School. In his spare time he did Chinese-language research for Human Rights Watch. Eventually, he became editor-in-chief of Columbia's *Journal of Asian Law.* He wore a certain expression I recognized from China—slightly stunned, a little overwhelmed, completely out of his element. He had no idea where this was going, but he was happy to hang on for the ride.

AT THE END OF THE DRIVE, WE FOLLOWED HIGHWAY 215 TO the Tibetan Plateau. The two-lane road was flanked by high desert landscapes of rock and dirt, punctuated by highway safety propaganda. Along one stretch, the government had perched a wrecked car on spindly ten-foot poles beside the road. The vehicle had been smashed beyond recognition; the front end was crumpled flat and the remains of a door dangled in strips of steel. Words had been painted across the back: "Four People Died." It was like some gruesome version of a children's treat— a Carsicle. Another sign presented the speed limit like options on a menu:

40 KM/HR IS THE SAFEST
80 KM/HR IS DANGEROUS
100 KM/HR IS BOUND FOR THE HOSPITAL

The road climbed steeply to the border of Qinghai Province. We passed slow-moving Liberation trucks, their engines whining; my altimeter read nearly twelve thousand feet. For 150 miles we saw almost no sign of human habitation. There were no gas stations or restaurants or shops; the first town we passed had been recently razed. Roofless walls stood stark on the plateau, lonely as the traces of some lost empire.

In Qinghai, Goettig's left eye began to act up. First it watered and then it hurt; he sat in the passenger's seat, rubbing his face with his fist. We crossed another twelve-thousand-foot pass and descended to Qinghai Lake. It's the largest lake in China, more than two hundred miles in circumference and blue as a sapphire. We camped on the banks of the salt lake, pitching my tent on a finger of land. It was one of the most beautiful places I had ever visited in China, but by now Goettig could hardly see a thing.

The next morning he lay in the tent, moaning. He had taken out his contact lenses, but the pain had increased; he asked how many hours it would be to Xining, the provincial capital. "It hurts like hell," he said. "It just keeps burning."

I asked if there was anything I could do.

"Maybe we'll have to find an eye doctor in Xining," he said. It occurred to me that this was the most ominous sentence I'd heard in about six thousand miles. The eye would eventually recover, and he later learned that the problem had been caused by his contacts. In Kunming, a friend had told him that a local shop was selling Johnson & Johnson lenses for half the usual price—a great deal, so Goettig stocked up. It turned out that the contacts were counterfeit. That became a new rule: when in Kunming, don't buy contact lenses on sale. China was full of lessons; we were still learning every day. Don't hike off trail in Xinjiang. Don't shop for Strange Stones in a bad part of Hebei. Don't hang out with people who light flares under stalled trucks. Driving along the lake, we passed another Carsicle, although Goettig's eyes were watering so badly he couldn't see it. He wept all the way across Qinghai—he wept along the salt lake's barren banks, and he wept past the stranded Carsicle, and he wept through the long descent from the roof of the world.

# ALL DUE RESPECT

ONE OF THE FOREMOST EXPERTS ON JAPANESE ORGANIZED crime is Jake Adelstein, who grew up on a farm in Missouri, worked as the only American on the crime beat for Japan's largest newspaper, and currently lives in central Tokyo under police protection. Japanese police protection means that the cops make daily visits to Adelstein's home, where they leave yellow notes pinned to the front gate that say, "There was nothing out of the ordinary." The notes feature the Tokyo police mascot, Pipo-kun, a smiling cartoon figure with big mouse ears and an antenna jutting out of its forehead. Some people in town have trouble taking Adelstein seriously. They dismiss him as a crank, a paranoid foreigner who talks obsessively about death threats from the gangsters known as yakuza. Others react with suspicion; a number of people in Japan claim that his journalism is a front for CIA work. There are Web sites that claim he is a Mossad agent. Adelstein does little to dismiss such rumors, apart from maintaining an image so flamboyant that it would shame any actual agency man. He's in his early forties, and he wears a trench coat and a porkpie hat, and he chain-smokes clove cigarettes from Indonesia. For a while, he dyed his hair bright red, claiming that this disguise would foil would-be assassins. He employs a bodyguard who doubles as a chauffeur, an ex-yakuza who cut off his pinky finger years ago as a gesture of apology to a gang superior. Adelstein says he needs a Mercedes and a nine-fingered driver in order to avoid the subway, where a hit man might shove him in front of a train.

Japan is not a dangerous country. Each year, approximately one murder is committed for every two hundred thousand people. This is among the lowest rates in the world, on a par with Iceland and Switzerland; the odds of being murdered in the United States are ten times higher. In Japan, it's a crime to own a gun, another crime to own a bullet, and a third crime to pull the trigger: three charges before you even think about a target. Yakuza are notoriously bad shots because practice is hard to come by, but somehow they have gained enormous influence. The police estimate that there are nearly eighty thousand members of yakuza organizations, whereas in America the Mafia had only five thousand in its heyday. The economic collapse of the 1990s is sometimes called "the yakuza recession," because organized crime played such a significant role.

"I can't think of a similar major civilized country where you have this kind of criminal influence," an American lawyer who handles risk assessment on behalf of a major financial firm told me recently, in Tokyo. He has a background in intelligence, and extensive experience reviewing potential investments to make sure they aren't connected to organized crime. "Every month, we turn away about a dozen companies that want to do business with us, because they have ties to the yakuza," he said. He told me that during the crash of 2008 Lehman Brothers lost $350 million in bad loans to yakuza front companies, while Citibank lost more than $700 million.

The lawyer didn't want me to use his name or identify his firm. "If you do this job correctly, and you're identifying bad guys and preventing business, then you're at risk," he said. He was familiar with Adelstein's work, and he noted that Adelstein took a completely different approach. "Jake has got a high profile," he said. "That's his style." He laughed about the clove cigarettes and the porkpie hat, but then he said, "If I were to learn that he was murdered this evening it wouldn't surprise me."

ADELSTEIN AND I BOTH GREW UP IN COLUMBIA, MISSOURI, and although I met him only a few times, he was the kind of kid that you don't forget. Back then, his name was Josh, and he was tall and thin, with an elongated face that seemed slightly lopsided. He was so cross-eyed that he had to have corrective surgery. Even after the procedure, his expression remained slightly off-kilter, and you could never tell exactly what he was looking at. Years later, he was given a diagnosis of Marfan syndrome, a rare disorder of connective tissue that often causes serious problems for the eyes, the heart, and other major organs. But as a boy he simply seemed odd. His vision and coordination were so poor that he didn't get a driver's license, an essential possession for any high school male in mid-Missouri, and he had to have classmates chauffeur him around town. He loved theater, which also qualified as a rare disorder in a sports-mad school; he was part of a circle that referred to itself as "the drama fags." Girls didn't have much use for him. The jocks teased and bullied him until a teacher suggested that he take up martial arts. Karate led to a freshman-year course in Japanese at the University of Missouri, which went well until Josh fell down an elevator shaft while working at a local bookstore. Even this was a sort of distinction—there aren't all that many elevators in Columbia, Missouri. Josh spent a week in the hospital with a bad head injury, and although he recovered, he couldn't remember any Japanese. But the head trauma also erased many memories of high school, so it may have been a good trade. He could always learn Japanese again.

He spent his sophomore year in Tokyo and never came back. He transferred to a Japanese university, and as a student he lived in a Zen Buddhist temple for three years. Somewhere along the way, he abandoned plans to become an actor, and he changed his name to Jake, for reasons that seemed to vary depending on when you asked him about it. He learned Japanese so quickly that within five years of studying the language he had passed the three-part exam to become a police reporter for Tokyo's *Yomiuri Shinbun*. Adelstein is believed to be the first American ever to make it through the newspaper's rigorous exam system.

The *Yomiuri* is the largest daily in the world. It prints two editions every day, and the total circulation is thirteen and a half million, more than ten times higher than that of the *New York Times*. The Internet has hardly affected the *Yomiuri*, which does not emphasize its Web site. Bylines are rare; most stories are covered by teams of journalists. At the *Yomiuri*, rookie police reporters are assigned to cover high school baseball, because the sport is supposed to be good training for crime journalists—the teamwork, the statistics, the attention to detail. When Adelstein started at the paper, his colleagues were stunned to discover that an American male could be totally ignorant about the sport. He didn't know the difference between an inning and an out; reading a box score was harder than staring at a page full of *kanji*. He told me that he spent his training period longing for a major crime to be committed. "In the middle of the high school baseball season, we were saved by the murder of this really beautiful girl who was killed and her body was found in a barrel," he said. "It's terrible to say, but I was happy to be doing something different."

In 2004, when I was living in China, I made a trip to Tokyo and contacted Adelstein. One evening, he gave me a tour of the red-light district in Kabukicho, telling outlandish stories about yakuza pimps. As part of his job, the *Yomiuri* provided a car and a full-time driver. Adelstein sat in back, dressed in a suit and tie; periodically he instructed the chauffeur to stop so he could meet a contact at a pachinko parlor or a dodgy massage joint. The last time I had seen him, a high school buddy was driving him around mid-Missouri in a station wagon, because his vision was so bad, but now he had transformed backseat status into a mark of prestige. A Missouri friend named Willoughby Johnson once said that Adelstein was still essentially an actor. "There's a degree to which anybody who becomes a character does so through self-fashioning," Johnson told me recently. He had been Adelstein's most faithful chauffeur in high school, and he still called him Josh. "I think of Josh in this way," he said. "He decided that he wanted to become this international man of mystery."

IN JAPAN, THE YAKUZA SOMETIMES SPEAK OF THEMSELVES IN terms of acting. "It's an atmosphere, a presence," an ex-gang member once told me. As a young criminal he had been given important advice by his *oyabun*, the "foster parent" within his gang. "My *oyabun* told me that when you're a yakuza, people are always watching you," he said. "Think of yourself as being onstage all the time. It's a performance. If you're bad at playing the role of a yakuza, then you're a bad yakuza, and you won't make a living."

The image has always been that of the underdog who survives through toughness and guile. The name refers to an unlucky hand at cards—*yakuza* means "eight-nine-three"—and bluffing is a big part of the routine. Many gangsters are Korean-Japanese or members of other ethnicities that traditionally have been scorned. These outsiders proved to be nimble after Japan's defeat in the Second World War, an era that is explored in *Tokyo Underworld*, by Robert Whiting. During this period, organized-crime groups established black markets where citizens could acquire necessities, and they were skilled at dealing with the occupying Americans. As Japan rebuilt itself, the yakuza got involved in real estate and in public-works projects.

For the most part, they eschewed violence against civilians, because the image of criminality was effective enough in an orderly society. Gangsters decorated their backs and arms with elaborate tattoos, and they permed their hair in tight curls that stood out among the Japanese. If a yakuza displeased a superior, he chopped off his own pinkie finger as a sign of apology. Gang members excelled at loan-sharking, extortion, and blackmail. They found creative ways to terrorize banks. In Tokyo, I accompanied Adelstein on a visit to the home of an aging midlevel gang member, who, along with a former colleague, reminisced about extorting banks in the 1980s.

"Sometimes we'd send three guys with cats, and they would twirl the cats around by the tail in front of the bank," one said, with Adelstein translating. "They'd do that until the bank finally gave them a loan. Or we'd have a hundred yakuza line up outside a bank. Each would go in and open an account for one yen, which was the lowest amount allowed for a new account. It would take all day, until finally the bank would agree to give some loans, to get rid of us." He said they wouldn't pay the loans back. "But we'd give the bank some protection, as well as help with collecting other bad loans," he said. "So it wasn't a terrible deal for them."

Both men had served time in prison for minor crimes. They were heavyset, with broad noses that looked to have been broken in the past. They had a cool, understated way of speaking, but their eyes were incredibly expressive—they had high arched brows, as fine as manga brushstrokes, that fluttered whenever they got excited. One had had his shoulders and arms tattooed with chrysanthemums, a patriotic symbol of imperial Japan. Each man said that he knew about a hundred colleagues who had died in various gang struggles. "It's part of yakuza life," one said. "You kill people, and eventually you get killed." But they emphasized that they hadn't targeted innocent civilians. They believed that true yakuza do honorable work: they go after deadbeats who don't repay loans, and they allow people to solve problems without wasting money on lawyers. Yakuza groups also engage in charity, especially after earthquakes or other disasters.

Many yakuza became rich during the bubble economy of the 1980s and 1990s, and they developed extensive corporate structures. (There's never been a law that bans the groups, which are fully registered.) Nowadays, yakuza run hedge funds. They speculate in real estate. The Inagawa-kai, one of the three biggest groups, keeps its main office across the street from the Ritz-Carlton Hotel in midtown Tokyo. At least one Japanese prime minister has been documented socializing with yakuza, and politicians have the kind of contact with criminal groups that would destroy a career elsewhere. In the mid-1990s, Shizuka Kamei, who was the minister of exports, admitted that he had accepted millions of dollars in donations from a yakuza front company, although he denied being aware of the criminal links. This did so little damage to his reputation that he eventually became minister of the agency that regulates Japan's finance industry.

AS A FOREIGNER, ADELSTEIN MOVES EASILY BETWEEN THE yakuza and the police, playing the flamboyant outsider with both. But he follows strict rules: information that comes from cops can be taken to other law-enforcement officials, but it cannot be passed to yakuza. In contrast, if a yakuza tells Adelstein something, the goal is usually to expose a rival group, so this information can be passed on to the cops. Adelstein is adamant about protecting sources. He says that the key to his work is the Japanese concept of *giri*, or reciprocity. His typical routine involves exchanging small favors with contacts, collecting bits of information that can be leveraged elsewhere.

One spring afternoon, I accompanied Adelstein to a Mexican restaurant in the neighborhood of Roppongi, to meet with a gangster who had a favor to ask. He was around forty years old—I'll call him Miyamoto—and he was college-educated, with perfect English. During his pre-yakuza days, he had worked for a public-relations firm in Tokyo. Back then, one of the agency's clients was an American auto manufacturer that regularly sent high-level management to Japan. In the evenings, it was Miyamoto's task to escort the gaijin to the massage parlors known as "soapland," where customers can enjoy a bath, a massage, and sex. Eventually, some yakuza extorted the public-relations firm, threatening to go to the tabloids with stories about American auto execs at soapland. Miyamoto handled the payoff, and then the next shakedown, and soon he became the firm's de facto yakuza liaison. The gangsters liked what they saw and recruited him away from the agency.

Since then, Miyamoto had become a full gang member, although you couldn't tell from looking at him. His *oyabun* had told him to avoid tattoos, because they would be a liability in the corporate world. He had kept all his fingers for the same reason. Nowadays he helped his gang manage three hedge funds. At the restaurant, he handed Adelstein a new business card. "Be really careful with this card, because it's my legitimate business," he said, in English. "If this gets out we won't get listed on the stock market."

The favor he needed was personal. His wife had left him after he became a yakuza, and he hadn't seen his child for years. The 2011 tsunami, which had occurred less than two months earlier, made him want to get back in touch. He asked Adelstein to contact his estranged wife. "Tell her I'm clean, and that I'm not a yakuza anymore," he said.

"I can't lie to her," Adelstein said. "I can say you're doing legitimate business. But I can't say you're not a yakuza."

"OK, I understand. Just try to convince her to see me again."

Miyamoto talked about other corporate yakuza, mentioning a well-known gang. "They now have a guy who worked for Deutsche Bank," he said.

Adelstein remarked that Miyamoto had posted his gang symbol online, and he warned him to be careful. "You need to back off on Twitter."

"Man, I've got a thousand followers!"

"You shouldn't say that stuff on Twitter about your bitches giving you money."

"The police won't read it. People think it's fake, anyway."

"Well, there's a new law going on the books in October, and if you're talking about taking protection money you could get arrested," Adelstein said.

"Yeah, I know."

There was never any mention of what Miyamoto might do in exchange for Adelstein's contacting his wife. But after a while the yakuza leaned forward and spoke in a low voice about the Tokyo Electric Power Company, or TEPCO, which owns and manages the Fukushima nuclear reactors that had been damaged by the tsunami. There had been accusations of mismanagement, and Miyamoto suggested that Adelstein research potential links between TEPCO and the Matsuba-kai, a criminal organization. "You know what's really interesting?" he said. "The Matsuba-kai guys play golf with the waste disposal guys for TEPCO. That's what you need to look into." He also named a yakuza from another gang who had supposedly made a million-dollar profit from supplying workers and construction materials to the reactors.

During the following weeks, Adelstein took pieces of information about the reactors to various contacts. Over the summer, he published a number of articles in *The Atlantic* online, the London *Independent*, and some Japanese publications, exposing criminal links at TEPCO. He described how yakuza front companies had supplied equipment and contract workers, and he quoted an engineer who had noticed something strange when he saw some cleanup crews change clothes: beneath the white hazmat suits, their bodies were covered with tattoos.

\* \* \*

WHEN ADELSTEIN WORKED FOR THE *Yomiuri*, HE SAYS THERE was a tacit understanding that investigative reporting on the yakuza shouldn't go too far. Media companies, like many big Japanese corporations, often had links to criminal groups, and even the police tried not to be too combative. For one thing, tools were limited: Japanese authorities can't engage in plea bargaining or witness relocation, and wire-tapping is almost never allowed. In the past, yakuza were rarely violent, and if they did attack somebody it was usually another gang member, which wasn't considered a problem. One officer in the organized-crime-prevention unit told me that in the 1980s, if a yakuza killed a rival he often turned himself in. "The guilty person would appear the next day at the police station with the gun and say, 'I did it,'" the cop said. "He'd be in jail for only two or three years. It wasn't like killing a real person."

Even the police officer believed that yakuza serve some useful functions. "Japanese society doesn't really have any place for juvenile delinquents," he said. "That's one role the yakuza play. Traditionally, it's a place where people can send juvenile delinquents." The fact that these delinquents are subsequently raised to become yakuza didn't seem to bother the cop too much. When I asked if he had ever fired his gun, he said that he hadn't even used his nightstick. His business card identified his specialty as "Violent Crime Investigation," and it featured the smiling Pipo-kun with his mouse ears and antenna, which symbolized how police can sense things happening everywhere in society. The officer explained that until recently the cops would notify yakuza before making a bust, out of respect, which allowed gangsters to hide any particularly damning evidence. "Now we don't do that anymore," he said.

He lamented a loss of civility among a new generation of yakuza. "It used to be that they didn't do theft or robbery," he said. "It was considered shameful. But now that's not the case anymore." He blamed greed: when the bubble economy collapsed in the 1990s, many wealthy yakuza had trouble adjusting. After years of adopting the facade of dangerous sociopaths, some began to live up to the image. The officer identified a gangster named Tadamasa Goto as an example of the new breed. "He's much more ruthless than yakuza were in the past," he said. "He'll go after civilians. Unfortunately, more yakuza have become like that."

Six days before our conversation, one of Goto's former underlings had been shot dead in Thailand. For years, he had been on the run, a suspect in the murder of a man who had stood in Goto's way in a real-estate deal. The cop said that Goto was cleaning up potential witnesses, and he reminded me that the gangster had also issued death threats against Adelstein. The most recent had been made last year, when Goto published his autobiography. "We suspect Goto of being involved in the killing of seventeen people," the cop said. "That murder in Thailand means that he can still reach out."

The criminal autobiography is a perverse genre anywhere in the world, but this is especially true in Japan, where Goto's book appeared with the title *Habakarinagara*, a polite phrase that means "with all due respect." At the time of publication, the author announced that all royalties would be dedicated to a charity for the disabled in Cambodia and to a Buddhist temple in Burma. The book begins in a David Copperfield vein: As a boy, Goto lacked shoes, and he had to eat barley instead of rice. ("Those years were extremely tough, with an alcoholic bum for a father.") He uses a nice baseball metaphor to describe the rise from juvenile delinquency to the yakuza. ("I felt as though we had been playing neighborhood baseball in a weedy field and suddenly got scouted to play in the major leagues.") Crimes are mentioned breezily, with few details, although even the offhand ones tend to be memorable. ("My third brother, Yasutaka, was

one of the guys who threw leaflets and excrement around Su-
ruga Bank, and he went to prison for that.") Goto emphasizes
his sense of honor; if nothing else, he has the courage of his con-
victions. ("I couldn't go apologize and beg forgiveness. I am not
cut out that way. I have pride. So instead I chopped off one of my
fingers and brought it to Kawauchi.")

For years this auto-amputee was one of the largest share-
holders of Japan Airlines. According to police estimates, Goto's assets are worth about a billion dollars, and he controlled
his own faction within the Yamaguchi-gumi, the top criminal
organization in the country. He is notorious for an attack on
Juzo Itami, one of Japan's greatest filmmakers. In May of 1992,
Itami released *Minbo, or the Gentle Art of Japanese Extortion*,
a movie that portrayed yakuza as fakes who don't live up to their
tough-guy image. Days later, five members of Goto's organization attacked the filmmaker in front of his home, slashing his
face and neck with knives. Goto claimed that he hadn't known
about the attack in advance. In his book, he describes it with
a mixture of surprise and pride, like a boss who returns from
vacation to find that his staff has been proactive. ("My first
thought was, 'If I find out whose men did this, I'm going to have
to send them a little something to pay my respects.'")

Afterward, Itami became even more outspoken. Five years
later, he apparently committed suicide, leaping from the roof
of his office building. He left a note explaining that he was dis-
traught over an alleged love affair. But Adelstein, citing an un-
named yakuza source, subsequently reported that the filmmaker
had been forced to sign the note and jump, and the police have
treated the case as a possible homicide. The American lawyer
who researches organized crime told me that some yakuza groups
specialize in murders that look like suicide. "I used to think they
committed suicide out of shame, because the Japanese do that,
culturally," he said. "But nowadays when I hear that somebody
killed himself I often doubt that's what happened."

At the *Yomiuri*, Adelstein started investigating Goto. He had been making progress when one of his sources, a foreign prostitute, disappeared. Adelstein was convinced that she had been murdered, and soon he became obsessed with the case. He was married to a Japanese journalist named Sunao, and they had two small children. But Adelstein rarely made it home before midnight, because Japanese crime reporters are expected to smoke and drink heavily with cops and other contacts. Sometimes he was threatened by yakuza; once he was badly beaten and suffered damage to his knee and spine. Like many people with Marfan syndrome, he took daily medication for his heart, and there were signs that his lifestyle was becoming self-destructive. He had always had a tendency to dramatize his health problems—this was part of his image—but now he seemed to be growing into the role of the troubled crime reporter.

Years later, both Adelstein and his wife said that this period destroyed their marriage. It also finished his career at the *Yomiuri*. After a certain point, he says, the paper balked at publishing more stories about Goto, and Adelstein quit. To this day, nobody at the paper will speak on the record about him; some reporters told me that he was a liar, while others said that the *Yomiuri* had been frustrated by his obsession. A couple of people alleged that he worked for the CIA. Staff from competing papers seemed more likely to praise his work, and a number of people indicated that the Japanese media tended to shy away from stories that would anger powerful yakuza figures like Goto. They also said that people at the *Yomiuri* were angry about Adelstein's departure because it violated traditional corporate loyalty.

After leaving the *Yomiuri*, Adelstein kept investigating, until finally he honed in on Goto's liver. For yakuza, the liver is a crucial body part, a target of self-abuse on a par with the pinkie finger. Many gangsters inject methamphetamines, and dirty needles can spread hepatitis C, which is also a risk of the big tattoos. In addition, there's a lot of drinking and smoking. In the yakuza community, a sick liver is a badge of honor, something that a proud samurai like Goto brags about in his memoirs. ("I drank enough to destroy three livers.") But it also means that yakuza often need transplants, and a criminal source told Adelstein that Goto had received a new liver in the United States, where his extensive record would make him ineligible for a visa. After months of investigation, Adelstein discovered that Goto and three other yakuza had been patients at the UCLA Medical Center, one of the nation's premier transplant facilities. Goto had been granted a visa because of a deal with the FBI—he had agreed to rat out other yakuza.

Adelstein broke the story in May of 2008, first in the *Washington Post* and then with details that he gave to reporters at the *Los Angeles Times*. Jim Stern, a retired head of the FBI's Asian criminal-enterprise unit, confirmed the deal, although he described it as a failure. He told the *LA Times*, "I don't think Goto gave the bureau anything of significance." (Stern was not involved in the deal.) According to Adelstein's Japanese police sources, the UCLA Medical Center received donations in excess of $1 million from yakuza. An investigation at UCLA found no wrongdoing, and the medical center reported only $200,000 in donations, although it acknowledged other signs of *giri*—Goto gave his doctor a case of wine, a watch, and $10,000. The year that he received his liver, 186 Americans in the Los Angeles area died while waiting for a transplant. Long before the articles appeared, Goto's men had contacted Adelstein. He says that they threatened to kill him, while another gang leader offered him half a million dollars to drop the story. After that, Adelstein was placed under Tokyo police protection, and the FBI monitored his wife and children, who had moved to the United States.

In 2008, the Yamaguchi-gumi officially expelled Goto. He undertook the training necessary to be certified as a Buddhist priest, a step that's not uncommon for ex-yakuza who fear retribution from former colleagues. It's bad karma to kill a priest, even if he's a former crime boss who reportedly still commands many loyal followers. And Goto is the type of Buddhist priest who uses his autobiography to issue oblique threats. "Just because I've retired from the business doesn't mean I have the time to track down this American novelist," he says in *With All Due Respect*. "If I did meet him, it would be a serious matter. He'd have to write, 'Goto is after me' instead of 'Goto may come after me.'"

In 2010, Adelstein hired a lawyer named Toshiro Igari to sue Goto's publisher and force the retraction of this threat. Igari was involved in many anti-yakuza cases, including investigations into fixing sumo matches and professional baseball games. In August of 2011, the lawyer went on vacation to the Philippines, where he was found dead in a room with a cup of sleeping pills, a set of box cutters, a glass of wine, and a shallow cut on his wrist. The Philippine police report was inconclusive, although most Japanese newspapers reported the death as a suicide. In Japan, Goto's book has sold more than two hundred thousand copies, and, since the spring of 2011, all royalties have been dedicated to tsunami relief.

DURING THE SPRING, I VISITED ADELSTEIN IN TOKYO, AND THE first thing he told me was that a week earlier he had been given a diagnosis of liver cancer. He had also nearly completed training to become a Zen Buddhist priest. Adelstein figured that if Goto could do it for protection, he could, too. He considered himself a Buddhist, and he liked the concept of karma, although he had told the priest who was training him that he didn't believe in reincarnation. "He said you don't have to believe," Adelstein said. "In Buddhism, it's not about faith; it's about doing."

He seemed neither surprised nor upset about the cancer diagnosis. The disease had been discovered in the early stages, and doctors at a clinic in Tokyo were treating it with injections of ethanol. They had told Adelstein that the cancer might be connected to diet, or to years of drinking and smoking, or even to Marfan syndrome. Regardless, his tranquility probably had less to do with Zen than it did with operating in a milieu where everybody knows something about liver problems. One afternoon, we stopped by the neighborhood police station, where Adelstein mentioned the diagnosis to a detective friend. "Wow, you're just like a yakuza!" the cop said. "Are you actually covered with tattoos?" When we met with one of Adelstein's criminal contacts, he talked about how his gang boss had originally hoped to get a UCLA liver, but after Adelstein's exposé he had been forced to settle for an Australian organ instead. (He eventually went through two Aussie livers, and then died.) Periodically, Adelstein's driver gave updates on a mutual acquaintance whose liver hadn't responded to ethanol and was currently being zapped with radio-wave treatment. The driver himself had a lucky liver—his hepatitis C had been successfully treated with interferon.

The driver's name was Teruo Mochizuki, and he had a long criminal history. As a teenager, he had been a delinquent, until finally his frustrated parents passed him off to a local yakuza. Mochizuki joined the Inagawa-kai, and he also became addicted to methamphetamines. He had gone to prison four times on drug-related charges. Now in his fifties, he said he had been clean for more than two decades. He was powerfully built, with broad shoulders, no neck, and a bullet-shaped head. Like other yakuza I met, he had expressive eyes, although even the manga brows remained still when I asked about his left hand. He said quietly, "There was some trouble and I had to lose the finger." He had done it in front of the boss at the gang's office. A doctor stopped the bleeding, but Mochizuki had declined treatment of the nerve endings. "To repair the finger would be to take back the apology," he explained. He said that the yakuza tradition is connected to the way that samurai warriors ritually sliced their own stomachs in ancient times. He also remarked that Japanese law grants disability status to a nine-fingered person, but Mochizuki refused to apply, out of respect for his digital apology.

He had known Adelstein for more than fifteen years. When I asked how they had first met, he told the story casually, as if these were the details of an everyday personal encounter. In 1993, an associate of Mochizuki's was blackmailing the criminal owner of a pet store, so the owner murdered the yakuza and, according to rumor, carved up the body and fed it to his dogs. Adelstein, who was single at the time, covered the story and interviewed the dead yakuza's meth-head girlfriend; almost immediately they began sleeping together. One day, Mochizuki went to the girlfriend's home to pay his condolences, and Adelstein answered the door, postcoital.

I had lots of possible questions but decided to go with the most obvious: "What was your first impression of Jake-san?"

"My first impression was, 'What an idiot!'" Mochizuki said. "You can look all over Japan and you won't find a reporter willing to do these things. I was surprised that he was fearless. He was just so strange."

Over the years, Adelstein and Mochizuki became friends. In 2007, Mochizuki was expelled from the Inagawa-kai, after an internal conflict that he didn't want to talk about. The following year, Adelstein offered Mochizuki a job as his bodyguard and driver. "I didn't want to do it," Mochizuki told me. "Goto is one of the most influential guys in Japan, and nobody would want a job like that. But I felt I had no other choice." He explained that Tokyo job prospects are poor for an uneducated middle-aged man with nine fingers and tattoos that show beneath a dress shirt. He now earns about $3,500 a month for driving Adelstein around Tokyo in a black Mercedes S600, which is a common yakuza car model. Adelstein had bought it cheap from another gang contact.

Mochizuki carried no weapons. He said that he tried to anticipate problems, and he monitored underworld contacts for news of Goto. He told me that Adelstein behaved differently from typical Japanese journalists, who are careful not to cross certain lines. "He has no regard for those taboos or restrictions," Mochizuki said. "If he were Japanese, he wouldn't be around right now." Mochizuki explained that some yakuza dislike Adelstein's stories, but he is widely recognized as a man of his word. "He has a heart," Mochizuki said. "People appreciate him for that. It's not common for somebody who is not Japanese to have this feeling of obligation."

Adelstein has published a book about his adventures on the police beat, *Tokyo Vice*, and is working on two more. A few years ago, he researched human trafficking for the U.S. State Department, and now he serves as a board member for the Polaris Project Japan, a nonprofit that combats the sex trade. Periodically, he does investigations for companies. The American lawyer who researches organized crime told me that when he first met Adelstein his image was off-putting. But he had become deeply impressed by his work. "He's a craftsman," he said. "He takes pride in doing the kind of research he does correctly." He continued, "It's this odd thing where you have this white guy who is as close to that part of Japanese society as a person can get."

Adelstein follows his strict rules of reciprocity and protection of sources, but otherwise he is willing to do nearly anything to get a story. He said that once, after his marriage had fallen apart, a lonely female cop offered access to a file on Goto if he slept with her, so he did. In the red-light district, he relies on foreign strippers for information, and on a few occasions when they've run into visa problems he has introduced them to gay salarymen who need wives in order to rise at their conservative Japanese companies. Adelstein says he never breaks the law—he simply puts these people in touch and tells them that they are free to fall in love and get married, and then they are also free to apply for spousal visas and show up at corporate events together. But he acknowledges that a journalist in America would be appalled. "I've slept with sources," he told me. "I've done hard negotiations that are probably tantamount to blackmail. I've ransacked rubbish bins for information. I'm willing to get information from organized crime or antisocial forces if the information is good."

By now, he's played the stereotypical role of the crime reporter for so long that he can't shake the lifestyle. Whenever I went out with him, we always seemed to end up having drinks with some beautiful, bright woman. For five years, he has rented a house in a quiet neighborhood, but it was as if he had just moved in: at night, he dragged a futon out of a closet and slept on the floor of his office. For breakfast he microwaved instant meals from a convenience store and served them on paper plates. In the kitchen, I counted five bottles of whiskey, four bottles of vodka, and three spoons. There was no table; he ate takeout meals on the couch. He marked his hand with a pen every time he lit a clove cigarette, supposedly to cut back, although once I watched him accumulate six marks while we were en route to a cancer treatment. On that particular day, the doctor decided to postpone the injection of ethanol, but I wasn't sure that Adelstein's body noticed the difference. We went straight from the appointment to dinner at a shabu-shabu restaurant, where he ordered two bottles of sake and finished them while waiting for an elegant Japanese-American woman to join us. After that, he had five more drinks at three different bars, and he was still going strong at two in the morning.

IN THE FARMLAND OF SOUTHERN BOONE COUNTY, ATOP THE last line of hills that overlook the Missouri River, stands a six-sided pagoda. The structure has three tiers marked by upturned eaves. "It was my impression of what a Japanese house should be," Eddie Adelstein told me, when I visited. He said that he didn't know much about Japan, having traveled there only once to see his son, but he had always liked the idea of an Asian house. He had a friend in Kansas who specialized in designing six-sided buildings, so they combined their interests. Since 2005, the pagoda has been home to Sunao Adelstein and her two children. Eddie and his wife, Willa, live in another building on the property.

The neighbors are mostly farmers and people who moved to the countryside for the quiet, but they've picked up certain ideas about the yakuza and Tadamasa Goto. "When it started, somebody from the FBI came by and talked to everybody," Heidi Branaugh, a nurse who lives on a small farm nearby, told me. "It was just odd. The first night I was here, after the sheriff came by, there was a helicopter overhead." Branaugh keeps a donkey, forty chickens, ten goats, and a dog named Bessie. She said that for a year Bessie barked at the car that the sheriff's department parked near the pagoda every evening. "They'd sit right up there on the drive, watching. Once they chased some guys who were looking for mushrooms." Another neighbor named Robert asked me for an update on Goto's threat. "It's safe now, isn't it?" he said. Beni Adelstein, Jake and Sunao's eleven-year-old daughter, wanted to make sure that I understood that yakuza are different from Missouri criminals. "It's not like they rob trains like Jesse James," she said.

Sunao Adelstein told me that she was tired of thinking about Goto. In 2008, the FBI had advised the family to install an alarm system and buy some guns, because it wouldn't be too hard for a hit man to track down the only pagoda along this stretch of the Missouri River. Since then, the authorities believed that the local risk had passed, although at the time of my visit Sunao had not returned to Japan for two years, because the Tokyo police were concerned about the threat in Goto's *With All Due Respect*. Sunao used to work as a business reporter in Tokyo, but now she was studying accounting and trying to adjust to life in rural Missouri. She liked the pagoda, although she complained that there was almost no closet space, because the designer had been so obsessed with the Japanese exterior. Who would have imagined that a pagoda needs closets? "Very often I think, why am I living here?" she said. "I grew up in Saitama. It's not a big city, but it's a suburb of Tokyo. I never dealt with ticks, with bugs. I hate ticks!"

Sunao is a slender, pretty woman, and she took me for a walk in the countryside with the children. She wore a short red skirt and black leggings; periodically she stopped to check for ticks. After years of living apart, she and Jake had finally decided to file for legal separation. They were still on good terms, but she spoke sadly about the marriage. In her opinion, her husband had changed after his research took him deep into the criminal underworld. "He was beaten by somebody, so he was wary. He was not goofy Jake anymore," she said. "He would use words the yakuza way." She continued, "It has to do with the facial expression, the way they speak. When he got angry he was like this. We argued once and he said, '*Omae niwa kankeine! Kono Bakayaro!*' I thought, Oh, he knows bad Japanese now."

She said that at one time she had hoped he would find a different career, but now she realized that it would never happen. Some of Adelstein's friends and family told me that he was addicted to the excitement, while others mentioned that he was too attached to the character that he had created. But beneath the chaotic personal life there was also something deeply moralistic about his outlook. He seemed to have more faith in *giri* than he did in any system of justice, and he could respect even a criminal as long as the man kept his word. "He expects people to be fair and honest," his father told me.

Eddie Adelstein said that his own experiences with crime had influenced his son. He worked as a pathologist at the VA hospital in Columbia, and had served as the county medical examiner for more than twenty years. In the early 1990s, patients at the VA suddenly began dying at a high rate, and there were rumors about a nurse named Richard Williams. Finally, Dr. Adelstein commissioned an epidemiologist to perform a study. "After three days, he said, 'You know, this guy is murdering people,'" Dr. Adelstein told me. "Most deaths were clustered between the hours of one and three in the morning, which was when that nurse was working. And he had taken care of eleven of the thirteen people who had died in the study." The study found that it was ten times more likely that a patient would die during one of Williams's shifts than under another nurse's care.

Some believed that the nurse might be killing patients by injecting codeine, but nobody knew for certain. When Adelstein approached hospital administrators, their first response was to hide the findings. "Everybody who took part in the cover-up was promoted, and everybody who tried to expose it was punished," he said. Williams eventually left the hospital, but the director gave him a letter of recommendation that helped him find a job at a rural nursing home. During Williams's first year at the nursing home, thirty-three patients died, whereas there had been only six deaths in the ten months before he started. Dr. Adelstein and others took the story to the FBI, Congress, and *ABC News*. The FBI investigated, but forensic results were incomplete, in part because labs were too busy with tests related to the O. J. Simpson trial. Williams was charged, but the case was dismissed when prosecutors could not determine the cause of death. At last report, he was living quietly in suburban St. Louis. He is suspected of murdering as many as forty-two people, many of them war veterans—more victims than are attributed to Ted Bundy, or John Wayne Gacy, or Tadamasa Goto.

All of that had happened while Jake was starting his career in Japan. "It made me extremely distrustful of everyone," he told me. "The biggest lesson I took was that even when you're in the right, when you're doing something good, you won't be rewarded." And it occurred to me that the darkest element of Adelstein's life wasn't the image he projected of the tormented reporter, or even the crazy yakuza stories. Beneath all the exoticism, it was actually the normalcy of crime that was most disturbing. Whether you're in Missouri or Tokyo, things aren't always what they seem—the nurse might be a murderer, and the gangster might run a hedge fund.

During one of my trips to Japan, I contacted Tadamasa Goto's publicist, who said that his client wasn't accepting interviews. So I got in touch with Tomohiko Suzuki, a journalist who has written for yakuza fanzines, which cover criminals as celebrities. Recently there had been rumors that Suzuki was channeling messages from Goto.

We met at a Tokyo coffeehouse. Suzuki wore blue work clothes and heavy boots, because he had just returned from a charity event in a town called Minamisoma, which was still suffering from the effects of the tsunami. In recent weeks, yakuza had been donating aid, and Goto had pitched in by sponsoring the day's event, which was called With All Due Respect. When I asked if any famous yakuza had attended, Suzuki named one and said, "He's the guy who stabbed the cult member in front of the media." I didn't pursue the details; by now I understood that the blandly offhand tone of such statements was basically the point.

Suzuki said that Adelstein's status as a foreigner had protected him from Goto. People in law enforcement and diplomatic circles had told me that they still took the threats seriously, but Suzuki said that their caution wasn't necessary anymore. "Those are the kinds of things that yakuza say all the time," he said. "It's kind of like saying 'Hello' for a yakuza."

A few months later, though, there were reports that Goto had become formally active again in organized crime. Not long after that, new laws went into effect that finally made it illegal to pay off yakuza. It was unclear how rigorously such regulations would be enforced, but they seemed to reflect a growing desire to control criminal groups. Suzuki hadn't said anything about Goto's plans during our meeting. He had visited the crime boss just a week earlier. "I didn't notice anything wrong with him—he looked very healthy," Suzuki said. "I think UCLA did a good job."

ON THE MORNING OF MY DEPARTURE, MOCHIZUKI DROVE Adelstein and me to Narita Airport. Adelstein had heard that somebody had recently smuggled a Marine-issued rifle through customs. "I have a contact in customs that I'll talk to about it," he said. "There might be a story." Afterward, he planned to go to a press conference downtown, and he was dressed in black suit pants, a pin-striped shirt, and a black trench coat with a red silk lining. He wore his porkpie hat. We had been on the road for a few minutes before he realized that he had forgotten his shoes. He laughed hysterically at his bulky house slippers and said that he'd have to buy a pair of loafers at an airport shop.

He was scheduled to undergo a chemotherapy treatment in about a week. At 7:25, he lit the day's first clove cigarette, and he chain-smoked during the long drive to the airport. On the way, Mochizuki asked Adelstein if he'd like to go on a beach vacation with him. "We should do this before one of us has bad health," the driver said. A couple of nights earlier, I had been in the car when Adelstein asked Mochizuki if he had ever killed a person. The driver paused, as if choosing his words carefully. "I've never killed anybody who wasn't a yakuza," he said at last, laughing.

Stories tended to tumble out of Adelstein, full of crazy yakuza details, and today he told a new one. He said that during his period of obsessive research, he had conducted an affair with one of Goto's mistresses. The gangster reportedly kept more than a dozen women in Tokyo and other cities, and Adelstein slept with one who gave him useful information. Eventually, he helped her escape Goto by introducing her to a gay salaryman who needed a wife and was about to be posted overseas. He said that the couple still shared an address in Europe and got along very well. I asked what the mistress was like.

"We had this lovely conversation once in bed," Adelstein said. "She said, 'Do you love me?' I said, 'No, but I like you.' She said, 'I like you, too, you're a lot of fun.' Then she said, 'Are you sleeping with me to get information about Goto?' I said, 'Pretty much. What about you?' She said, 'Well, I hate the motherfucker and every time I sleep with you it's like I'm stabbing him in the face.' She was into astronomy. Once we went to a planetarium in Sunshine City. I think that was the only time we ever went out in public together. That was our only date." He continued, "It was nice. Another time I gave her a gift—I bought an expensive planetarium set that Sega makes. She cried."

He lit another cigarette. He had told me once that he didn't expect to have a long life, but in Tokyo he always seemed happy and full of energy. And I liked the image from his story: the odd couple at the planetarium, the Japanese gangster's mistress and the cross-eyed kid from Missouri, both of them staring up at the stars. I thought about that until we reached the airport and he went off to find some shoes.

# WHEN YOU GROW UP

LITTLE LU, LITTLE ZHANG, AND LITTLE LIU WAITED FOR ME at the end of the bridge. They were ten, twelve, and fourteen years old, respectively, and they had come from the same village in northern Sichuan Province. They said that they had dropped out of school and migrated to the south because their families were too poor to afford the school fees. I had met them three days earlier in downtown Shenzhen, where they had tried to sell me pornographic video disks.

They told me that at first they had worked for a man who hired children to sell pornography because they were too young to be sent to jail. He paid each child three hundred yuan a month, about thirty-six dollars. But the boys said that after a while they had gone freelance. Initially I had trouble believing this—it seemed impossible that children so young would be capable of handling an illegal business on their own. But during the month that I spent in Shenzhen, I visited them on a regular basis, and I never saw any sign of adult supervision. Eventually I came to believe that most of the things they told me were true. They claimed that twice they had been arrested and deported from the Shenzhen Special Economic Zone, but both times they had returned by climbing the chain-link fence that surrounded the city. They rented an apartment and cooked for themselves.

They slept in one bed. They bought pornographic disks for four yuan and sold them for ten, or a little more than a dollar. They pooled their money and earned enough for each boy to send at least three hundred yuan per month to his family.

I had promised to take them to lunch today. We found a Cantonese restaurant, where the boys sat down and immediately ordered iced coffee and hotpot, a combination that I had not enjoyed previously. After the hotpot arrived, the boys took every condiment container on the table and emptied them into the bowl. The hotpot oil bubbled around big clumps of salt, MSG, and hot pepper. They had done the same thing to fried chicken when I took them to KFC a few days earlier.

Within two minutes they had finished the iced coffee.

"I want a beer," said Little Lu. He was the youngest of the trio, but nevertheless he was the leader. I told him that he was too young for beer and that we would have tea instead.

They ate hungrily for a while and then Little Lu called for the waitress. I had never seen a ten-year-old speak to an adult with such authority.

"Give me a beer," he said.

"Don't give him a beer," I said. "We'll drink tea."

"I want a beer," said Little Lu. The waitress seemed unsure who was in charge here.

"No beer," I said firmly.

"You have to drink beer with hotpot," said Little Lu. "We do that at my hometown."

"His father has a big alcohol tolerance," said Little Zhang, pointing at Little Lu.

This was one of the things that I believed to be true, although I didn't want to pursue the subject right now. Instead, I asked about how they avoided getting caught by the police. Little Liu and Little Zhang said that they kept their hair very short so the cops had nothing to grab onto, and they avoided long-sleeved shirts for the same reason. Like the others, Little Lu was dressed in a tight short-sleeved shirt, but his hair was longer. It was parted carefully down the middle and he seemed vain about it. In the middle of the meal, he got up to use the bathroom; when he returned, his hair was slicked back. I had been watching him carefully and that was the only time that he was out of my sight. Almost immediately another waitress came over.

"Do you want Tsingtao or Yanjing?" she said.

"Don't bring us any beer."

"But he just ordered it!"

"Don't listen to him."

The boys finished the vegetables and meat in the hotpot, and then they slurped down the broth, which had acquired a bright chemical color from all the condiments. They still seemed famished. I asked what they wanted to be when they grew up.

"A driver," said Little Lu.

"A security guard," said Little Liu.

Little Zhang smiled and said, "I want to go home."

# QUARTET

THE FIRST ACCIDENT WASN'T MY FAULT. I HAD RENTED A Volkswagen Jetta and driven to my weekend home in Sancha, a village north of Beijing. I parked at the end of the road, where the pavement widens into an empty lot. It's impossible to drive within Sancha; like virtually all Chinese villages, it was built before anybody had cars, and homes are linked by narrow footpaths.

About an hour after I arrived, my neighbor asked me to move the car, because the villagers were about to mix cement in the lot. That day, Leslie, my wife, and I were both on our computers, trying to do some writing.

"I can move it if you want," my neighbor said. His name is Wei Ziqi, and he had recently completed a driving course and received his license. It was his proudest achievement—he was one of the first in the village to learn to drive. I handed him the keys and sat back down at my computer. Half an hour later, he returned and stood in the doorway silently. I asked if everything was all right.

"There's a problem with the car," he said slowly. He was smiling, but it was a tight Chinese grin of embarrassment, the kind of expression that makes your pulse quicken.

"What kind of problem?" I said.

"I think you should come see it."

In the lot, a couple of villagers were staring at the car; they were grinning, too. The front bumper had been knocked completely off. It lay on the road, leaving the Jetta's grill gaping, like a child who's lost three teeth and can't stop smiling. Why did everybody look so goddam happy?

"I forgot about the front end," Wei Ziqi said.

"What do you mean?" I said.

"I'm not used to driving something with a front end," he said. "During my course we only drove Liberation trucks. They're flat in front."

I had parked the Jetta parallel to a wall, and he had backed up and turned the wheel sharply, not realizing that the front end would swing in the opposite direction. I knelt down and inspected the bumper—it was hopelessly bent.

"How much do you think it's going to cost?" he said.

"I have no idea," I said. "I've never done something like this before."

He got some wire and tied the bumper to the front end. He offered repeatedly to pay for it, but I told him not to worry; I'd deal with the rental company. The next day, I set off to return the car.

DRIVING IS SOMETHING THAT I TAKE VERY SERIOUSLY. WHEN I turned sixteen, I was told that handling an automobile is a privilege and a responsibility, and I still get nervous thinking about the day that my mother drove me to the Wilkes Boulevard United Methodist Church, in Columbia, Missouri, to take my first driving exam. The state's Division of Motor Vehicles rented office space in the building, and the exam began and ended in the church parking lot. In mid-Missouri, it was widely known that when it came to judging sixteen-year-old males the DMV was even tougher than the Methodists. They failed boys for not checking the blind spot, for running yellow lights, for tiny adjustments on parallel parking. There were rumors that any boy who was visibly confident would flunk—if you believed that you were predestined for a license, then the folks at the Wilkes Boulevard United Methodist Church would prove you wrong. I took the exam in my family's Dodge Caravan, and afterward the examiner gave me a stern speech. It began with the statement "You're lucky we don't professionally evaluate you" and ended with "I hope I don't see you in the hospital someday." Between these remarks, the man acknowledged that I had passed by the barest of margins, which was all that mattered. There was no purgatory at the DMV. You either failed or you passed, and success meant that as long as you avoided trouble and kept up the paperwork, you'd never have to take another driving exam in the state of Missouri.

After moving to Beijing, I was surprised that my Missouri license had some currency in the People's Republic. The country was in the early stages of an auto boom; Beijing alone was registering almost a thousand new drivers every day. All Chinese applicants were required to have a medical checkup, take a written exam, enroll in a month-long technical course, and then pass two driving tests. But the process had been pared down for any foreigner who already had certification from his home country, and all I had to do was pass a special foreigner's road test. The examiner was in his midforties, and he wore white cotton driving gloves with tobacco stains on the fingers. He lit up a Red Pagoda Mountain as soon as I got in the car. It was a Volkswagen Santana, the nation's most popular passenger vehicle at the time.

"Start the car," the man said, and I turned the key. "Drive forward," he said.

We were north of the city, in a neighborhood that had been cleared of all traffic—no cars, no bikes, no pedestrians. It was the most peaceful street I'd ever seen in the capital, and I wish I could have savored it. But after fifty yards the examiner spoke again. "Pull over," he said. "Turn off the car."

The Santana fell silent; the man filled out forms, his pen moving efficiently. He had barely burned through the tip of his Red Pagoda Mountain. "Is that all?" I said.

"That's it," the man said. He asked me where I had learned Chinese, and we chatted for a while. One of the last things he said to me was, "You're a very good driver."

That summer, I began renting cars from a company called Capital Motors. The car-rental industry was a new one; five years earlier, almost nobody in Beijing would have thought of hiring an automobile for a weekend trip. But now my local company had a fleet of about fifty vehicles, mostly Chinese-made Jettas and Santanas. Usually, I rented a Jetta, which cost twenty-five dollars per day and involved an enormous amount of paperwork. The most elaborate part of the process was a survey of the car's exterior, led by an employee, who recorded dents and scratches on a diagram. This inspection often took a while—the Jetta is a small automobile, but the marks of Beijing traffic made the most of the limited canvas. After documenting the damage, the employee turned the key in the ignition and showed me the gas gauge. Sometimes it was half full; sometimes there was a quarter tank. Sometimes he studied it and announced: "Three-eighths." It was my responsibility to return the car with exactly the same amount of fuel. One day, I decided to make a contribution to the fledgling industry.

"You should rent cars with a full tank, and then require the customer to bring it back full," I said. "That's how rental companies do it in America."

"That would never work here," said Mr. Wang, who usually handled my rental. He was a big man with thinning hair that flopped loosely over a wide forehead; he always seemed to be in a good mood. He sat with two other men in the Capital Motors front office, where they smoked cigarettes as if it were a competition. The room was so full of smoke that I could barely read the company evaluation sign that hung on the wall:

CUSTOMER SATISFACTION RATING: 90%
EFFICIENCY RATING: 97%
APPROPRIATE SERVICE DICTION RATING: 98%
SERVICE ATTITUDE RATING: 99%

"That might work in America, but it wouldn't work here," Mr. Wang continued. "People in China would return the car empty."

"Then you charge them a lot extra to refill it," I said. "Make it a standard rule. Charge extra if people don't obey and they'll learn to follow it."

"Chinese people would never do that!"

"I'm sure they would," I said.

"You don't understand Chinese people!" Mr. Wang said, laughing, and the other men nodded. As a foreigner, I often heard that statement, and it had a way of ending discussion. The Chinese people had invented the compass, silk, paper, gunpowder, the seismoscope; they had sailed to Africa in the fifteenth century; they had built the Great Wall; over the past decade they had expanded their economy at a rate never before seen in the developing world. They could return a rental car with exactly three-eighths of a tank of gas, but filling it was apparently beyond the realm of possibility. We had a couple more conversations about this, but finally I dropped the subject. It was impossible to argue with somebody as friendly as Mr. Wang.

He seemed especially cheerful when I returned the Jetta with the ruined bumper. In the past, I had brought back cars with new dents; this was inevitable in a city with more than two million cars, most of them handled by rookies. But I had never done any serious damage, and Mr. Wang's eyes grew wide when he saw the Jetta. "Waah!" the man said. "How did you do that?"

"I didn't," I said. I described Wei Ziqi's lack of experience with hooded cars, and Mr. Liu looked confused; the more I expanded on this topic, the blanker his expression became. At last, I abandoned the front end—I offered to pay for the bumper.

"*Mei wenti!*" Mr. Wang said, smiling. "No problem! We have insurance! You just need to write an accident report. Do you have your chop?"

I told Mr. Wang that my chop—an official stamp registered to one's work unit, in my case, *The New Yorker*—was at home.

"No problem! Just bring it next time." He opened a drawer and pulled out a stack of papers; each was blank except for a red stamp. Mr. Wang rifled through the pile, selected a paper, and laid it in front of me. The chop read: "U.S.-China Tractor Association."

"What's this?" I said.

"It doesn't matter," he said. "They had an accident, but they didn't have their chop, so they used somebody else's. Then they brought this page to replace it. Now you can write your report on their page, and next time bring a piece of paper with your chop, so the next person can use it. Understand?"

I didn't—he had to explain this arrangement three times. Finally it dawned on me that the wrecked bumper, which had never been my fault, and in a sense wasn't Wei Ziqi's fault, either, because of the unexpected front end, would now be blamed on the U.S.-China Tractor Association. "But you shouldn't say it happened in the countryside," Mr. Wang said. "That's too complicated. Just say you had an accident in our parking lot."

He wrote out a sample report and Leslie copied it, because her written Chinese was much better than mine. I signed my name across the tractor chop. The next time I rented a car, Mr. Wang told me that the insurance had covered everything. He never hassled me about bringing in a chopped paper, and I decided to leave it at that—I was an old customer, as Mr. Wang liked to say.

\*    \*    \*

WHEN YOU LIVE IN CHINA AS A FOREIGNER, THERE ARE TWO critical moments of recognition. The first occurs immediately upon arrival, when you are confronted with your own ignorance. Language, customs, history—all of it has to be learned, and the task seems impossible. Then, just as you begin to catch on, you realize that everybody else feels pretty much the same way. The place changes too fast; nobody in China has the luxury of being confident in his knowledge. Who shows a peasant how to find a factory job? How does a former Maoist learn to start a business? Who has the slightest clue how to run a car rental agency? Everything is figured out on the fly; the people are masters at improvisation. This is the second moment of recognition, and it's even more frightening than the first. Awareness of your own ignorance is a lonely feeling, but there's little consolation in sharing it with 1.3 billion neighbors.

On the road it's particularly horrifying. China still doesn't have many drivers—when I got my license, there were only twenty-eight automobiles per every thousand people, which is about the same rate that the United States had in 1915. But a 2004 World Health Organization report found that China, while having only 3 percent of the world's vehicles, accounted for 21 percent of its traffic fatalities. It's a nation of new drivers, and the transition has been so rapid that many road patterns come directly from pedestrian life—people drive the way they walk. They like to move in packs, and they tailgate whenever possible. They rarely use turn signals. If they miss an exit on a highway, they simply pull onto the shoulder, shift into reverse, and get it right the second time. After years of long queues, Chinese people have learned to be ruthless about cutting in line, an instinct that is disastrous in traffic jams. Toll booths are hazardous for the same reason. Drivers rarely check their rearview mirrors, perhaps because they never use such an instrument when they travel on foot or by bicycle. Windshield wipers are considered a distraction, and so are headlights.

In fact, the use of headlights was banned in Beijing until the mid-1980s, when Chinese officials began going overseas in increasing numbers. These trips were encouraged by governments in Europe and the United States, in the hope that glimpses of democracy would encourage China's leaders to rethink their policies. In 1983, Chen Xitong, the mayor of Beijing, made one such visit to New York. During his meetings with Mayor Ed Koch, Chen made a crucial observation: Manhattan drivers turn on their lights at night. When Chen returned to China, he decreed that Beijing motorists do the same. It's unclear what political conclusions Chen drew from his encounters with American democracy—the man ended up in prison for corruption—but at least he did his part for traffic safety. Nevertheless, there's still enough debate about headlight use to merit a question on the written driving exam:

> 278. *During the evening, a driver should*
>    a) *turn on the brights.*
>    b) *turn on the normal lights.*
>    c) *turn off the lights.*

Recently, I picked up a study booklet for the exam. It consisted of 429 multiple-choice questions and 256 true-false queries, any of which might appear on the test. Often these questions successfully captured the spirit of the road ("True or False: In a taxi, it's fine to carry a small amount of explosive material"), but I wasn't convinced that they helped people learn to drive correctly. After carefully studying the document, though, I realized that it was descriptive rather than prescriptive. It didn't teach people how to drive; it taught you how people drove.

77. *When overtaking another car, a driver should pass*
    *a) on the left.*
    *b) on the right.*
    *c) wherever, depending on the situation.*

354. *If you are driving past a big puddle and there are pedestrians next to the water, you should*
    *a) accelerate.*
    *b) slow down and make sure that the water does not splash them.*
    *c) continue at the same speed straight through the puddle.*

80. *If, while preparing to pass a car, you notice that it is turning left, making a U-turn, or passing another vehicle, you should*
    *a) pass on the right.*
    *b) do not pass.*
    *c) honk, accelerate, and pass on the left.*

Lots of answers involved honking. In Chinese automobiles, the horn is essentially neurological—it channels the driver's reflexes. People honk constantly, and at first all horns sound the same, but over time you learn to distinguish variations and interpret them correctly. In this sense, it's as complicated as the language. Spoken Chinese is tonal, which means that a single syllable can have different meanings depending on whether it is flat, rising, falling and rising, or falling sharply. Similarly, a Chinese horn is capable of at least ten distinct meanings. A solid *hooooonnnnkkkkk* is intended to attract attention. A double sound—*hooooonnnnkkkkk, hooooonnnnkkkkk*—indicates irritation. There's a particularly long *hooooooooonnnnnnnnnkkkkkk* which means that the driver is stuck in traffic, has exhausted curb-sneaking options, and would like everybody else on the road to disappear. A responding *hooooooooooooonnnnnnnnnnnnnnnkkkkkkkkkkkk* proves that they aren't going anywhere. There's a stuttering, staggering *honk honk hnk hnk hnk hnk hnk hnk* that represents pure panic. There's the afterthought *honk*—the one that rookie drivers make if they are too slow to hit the button before a situation resolves itself. And there's a short, simple *honk* that says, "Nothing actually happened, but my hands are still on the wheel, and this horn continues to serve as an extension of my nervous system." Other honks can be found on the exam:

> 353. *When passing an elderly person or a child, you should*
> *a) slow down and make sure you pass safely.*
> *b) continue at the same speed.*
> *c) honk the horn to tell them to watch out.*

> 269. *When you enter a tunnel, you should*
> *a) honk and accelerate.*
> *b) slow down and turn on your lights.*
> *c) honk and maintain speed.*

> 355. *When driving through a residential area, you should*
> *a) honk like normal.*
> *b) honk more than normal, in order to alert residents.*
> *c) avoid honking, in order to avoid disturbing residents.*

THE SECOND ACCIDENT WASN'T MY FAULT, EITHER. I WAS DRIVING in the countryside, and a dog darted out from behind a house and lunged at my Jetta. This is a common problem—dogs, like everybody else in China, aren't quite accustomed to having automobiles around. I swerved, but it was too late; the dog thudded against the front of the car. When Leslie and I returned the Jetta, the three men were smoking cigarettes beside the company evaluation sign. Everybody at Capital Motors was still doing a bang-up job:

CUSTOMER SATISFACTION RATING: 90%
EFFICIENCY RATING: 97%
APPROPRIATE SERVICE DICTION RATING: 98%
SERVICE ATTITUDE RATING: 99%

Mr. Wang inspected the Jetta and noted cheerfully that the plastic cover for the right signal light had been smashed. He asked what I had hit.

"A dog," I said.

"*Gou mei wenti?*" he said. "The dog didn't have a problem, did it?"

"The dog had a problem," I said. "It died."

Mr. Wang's smile got bigger. "Did you eat it?"

I couldn't tell if he was joking—he was a dog-owner himself, and I had seen him playing with his pet in the office. "It wasn't that kind of dog," I said. "It was one of those tiny little dogs."

"Well, sometimes if a driver hits a big dog, he just throws it in the trunk, takes it home, and cooks it," he said. He charged us twelve dollars for a new signal-light cover—it was too minor for the insurance, and there was no need to call in the U.S.-China Tractor Association.

EVERY CHINESE APPLICANT FOR A LICENSE MUST ENROLL IN a certified course, at his own expense, and he must spend at least fifty-eight hours in training. This suggests a high degree of standardization, but much depends on the instructor, who is called a *jiaolian*, or "coach." Often, coaches have developed their own theories and regimens, like the martial-arts masters of old. Wei Ziqi's coach disdained front-end vehicles, and he also forced his students to begin every maneuver in second gear. It was more challenging, he said; first gear would only make them lazy. Another woman I know had a coach who forbade the use of turn signals, because they distracted other drivers. When Leslie decided to learn to drive stick, she hired a private instructor in Beijing. I had doubts about whether this would be effective, but I knew who the most obvious alternative coach would be, so I held my tongue. On Leslie's first lesson, the coach introduced himself, sat in the passenger seat, and adjusted the rearview mirror so that it faced him.

"How am I going to see what's behind me?" Leslie asked.

"I'll tell you what's behind you," the coach said. "You don't need to worry about that." He was like the martial-arts guru who blindfolds his pupil: trust is the first step toward mastery.

Recently, I went to observe some courses at the Public Safety Driving School in the southeastern city of Lishui. Local car ownership was still low—only twenty households out of every thousand had bought a car within the past six months. But that was twice the previous year's rate, and the city's factory economy was in the middle of a boom. The driving school was busy, and classes moved through three stages: the parking range, the driving range, and the road.

One afternoon, I watched six students embark on their first day. An instructor called Coach Tang began by raising the hood of a red Santana. He pointed out the engine, the radiator, the battery. He showed them how to unscrew the gas cap. The door was next—the students practiced opening and closing it. Then he identified the panel instruments and the pedals. The students circled the Santana warily, fiddling with parts, like the blind men and the elephant. Finally, after an hour, they were allowed to enter the vehicle. Each of them sat in the driver's seat, where they shifted repeatedly from first to fifth gear, with the engine off. Watching this made me wince, and after a while I said to Coach Tang, "Isn't that bad for the car?"

"No," he said. "It's fine."

"I think it might be bad if the motor's off," I said.

"It's completely fine," Coach Tang said. "We do it all the time." In China, instructors of any type are traditionally respected without question, and I decided to keep my mouth shut. But it wasn't easy. For the next step, the students learned to use the clutch by setting the parking brake, starting the engine, shifting into first gear, and then releasing the clutch while adding gas. The motor whined against the force of the brake; the torque dipped the front end up and down. By the end of the day, you could have fried an egg on the Santana's hood, and my palms began to sweat every time another driver gunned the engine. I could practically hear my father's voice—he's a good amateur mechanic, and few things anger him more than mindless abuse of an automobile.

Nobody was allowed to operate the vehicle until the second day of class. There were four men and two women, and all of them were younger than forty. Each had paid more than three hundred dollars for the course—a lot of money in a city where the monthly minimum wage was roughly seventy-five dollars. Only one person came from a household that currently owned an automobile. The others told me that someday they might buy one, and the university students—there were four of them— believed that a driver's license would look good on a résumé. "It's something you should be able to do, like swimming," a student named Wang Yanheng told me. "In the future, so many people in China are going to have cars." He was a senior, majoring in information technology. The only person from a home with automobiles (three) was a nineteen-year-old sociology major whose father owned a plastics factory. When I asked what the factory produced, the woman ran a finger along the rubber lining of the Santana's window. "This is one of the things we make," she said.

The students spent ten days on the parking range, and during that time they performed exactly three movements: a ninety-degree turn into a parking spot, the same maneuver in reverse, and parallel parking. Every day, for as many as six hours, they practiced these turns over and over. Like any good martial arts master, Coach Tang was strict. "What are you doing?" he yelled, when one student brushed against a pole. "You must have forgotten your brain today!" "Don't hold the gearshift loosely like that!" he shouted at another man. "If you do, your father will curse you!" Sometimes he slapped a student's hand. There was a strict rule against head turns—even in reverse, you were supposed to rely on mirrors only.

The next step was the driving range, where the skill set became more demanding. Drivers were required to stop within twenty-five centimeters of a painted line, and they followed an obstacle course of tight turns. The final skill was the "single-plank bridge"—a concrete riser, a foot high and only slightly wider than a tire. Students had to aim the car perfectly, so that two wheels perched atop the riser—first the left tires, then the right. If a single wheel slipped, they failed the exam. The students spent most of their ten days practicing the single-plank bridge, and I asked a coach why it was so important. "Because it's very difficult," he said.

"Right, I understand that," I said. "But when is it useful on the road?"

"Well, if you're crossing a bridge with a hole, and there's only one place where the tires can go, then it's important to be able to do this."

The Chinese have fantastic driving imaginations—the written exam was full of situations like this. They seemed ridiculously unlikely, but the level of detail was such that I suspected it must have happened to somebody, somewhere:

279. *If your car breaks down atop the tracks of a railroad crossing, you should*
   *a) abandon it there.*
   *b) find some way to move it immediately.*
   *c) leave it there temporarily until you can get somebody to repair it.*

The course ended with a week and a half on the road, and I

accompanied another class on their final day. With the coach in the passenger seat, students took turns driving along a two-lane rural road. There were certain movements they had to perform: shift to fifth, downshift to first, make a U-turn, stop at an imitation traffic light. They had been instructed to honk whenever they pulled out, or made a turn, or encountered anything in the road. They honked at cars, tractors, and donkey carts. They honked at every single pedestrian. Sometimes they passed another car from the driving school, and then both vehicles honked happily, as if greeting an old friend. At noon, the class had lunch at a local restaurant, where everybody drank beer, including the coach, and then they continued driving. One student told me that a day earlier they got so drunk that they had to cancel the afternoon class.

Throughout the course, there had been no variables, no emphasis on responding to situations. Instead, students learned and rehearsed a small number of set pieces, which they would later combine and apply to actual city driving. It reminded me of how Chinese schoolchildren learn to write: they begin with specific strokes, copying them over and over, and then they combine these into characters, which are also written repeatedly. In China, repetition is the cornerstone of education, and virtually every new skill is approached in this manner. It's one reason the Chinese have been far more successful at building assembly-line factories than at innovation.

It also explains some of the problems with driving in China. On the final day of class, a student begged me to let him drive my rental car back to the road range, for more practice. In a moment of extremely poor judgment, I agreed, and those turned out to be the most terrifying seven miles I had ever experienced in China. Twice I had to yell to keep him from passing on blind turns; another time, I grabbed the wheel to prevent him from veering into a car. He never checked the rearview mirror; he honked at everything that moved. The total absence of turn signals was the least of our problems. He came within inches of hitting a parked tractor, and he almost nailed a cement wall. When we finally made it to the range, I could have fallen on my knees and kissed the single-plank bridge.

Foreigners in Beijing often said to me, "I can't believe you're driving in this country." To which I responded, "I can't believe you get into cabs and buses driven by graduates of Chinese driving courses." Out on the road, everybody was lost—*une génération perdue*—but it felt better to be the one behind the wheel.

I HAD NOTHING TO DO WITH THE THIRD ACCIDENT. I COULDN'T even drive—I had broken my left kneecap while hiking on the Great Wall, and the Jetta that we had rented was standard transmission. Despite having served as a blindfolded acolyte of a local driving master, Leslie still didn't feel comfortable behind the wheel, and one afternoon she asked me to accompany her on some errands. I sat in the back, my broken leg propped up, giving advice every time she stalled. ("More gas!") It was snowing; traffic was miserable; we spent two hours hustling in and out of shops. After the last stop, Leslie turned the key and the Jetta lurched straight ahead into a brick wall.

I said, "Use the clutch."

There had been a distinct crunching sound, but we didn't check the car; by now we were desperate to get home. Near the Lama Temple, as we waited to make the last left turn of the day, we were hit by another car. The driver backed into our side and then pulled away. There wasn't time to fumble with the crutches, so I hopped out on my good leg. Fortunately, traffic was backed up, and I caught him in about seven hops. I pounded on the window. "You hit my car!"

The driver looked up, surprised: a one-legged foreigner, hopping mad and smacking the glass. He stepped out and apologized, saying that he hadn't felt the impact. Together we inspected the Jetta—fresh dent above the left rear wheel. The man said, "I'll give you one hundred." That was about thirteen dollars.

In China, after a minor accident, people usually settle the matter on the street, in cash. This routine has become a standard part of life—once I saw two small children playing a game in which they repeatedly rammed bikes and shouted, "*Pei qian! Pei qian!*"—"Compensate! Compensate!"

Leslie used her cell phone to call Capital Motors. Mr. Wang didn't sound the least bit surprised to hear that we'd had another accident. All he said was, "Ask for two hundred."

"That's too much," the other driver said. "This is really minor."

"It's not our decision."

"Well, then we'll have to call the police," he said, but it was clear that he didn't want to do this. A dozen bystanders had gathered around the cars, which were parked in the middle of the snowy street. With Chinese accidents, the crowd is more like a jury than an audience, and a middle-aged woman bent over to inspect the dent. She stood up and announced, "A hundred is enough."

"What do you have to do with it?" Leslie snapped. "You can't even drive!"

This may have been a case of the wok calling the kettle black, but I didn't say anything. And Leslie must have been right, because the woman shut up. But the driver refused to pay two hundred. "Should we accept one-fifty?" Leslie asked me, in English. Lao-Tzu said it best: A man standing on crutches in the snow will not bargain long over a dent to a crappy Jetta rental. Later that day, Leslie returned the car and the cash. Mr. Wang noticed that another light cover had been broken when she hit the brick wall. He said, happily, "What did you kill this time?" When I hit the dog, the same cover was twelve dollars; this time he asked for only three. It must have been a special price because we did so well at the Lama Temple.

※    ※    ※

THE FOURTH ACCIDENT WAS ENTIRELY MY FAULT. IT WAS MY last day in China, my last Jetta—the next morning I had a one-way ticket to Honolulu. On my way to return the car, I got stuck in a terrible traffic jam, and wailing horns filled the air—these were the honks that mean, "Let me out of here!" In front, a taxi-driver saw an opening and lurched ahead; I lurched after him; he stopped short; I didn't.

We got out. I took a look and winced: dents on both sides. "One hundred," I said.

"Are you kidding?" the man yelled. "This is at least two hundred!"

Suddenly, I felt extremely tired. Ten years in China, six years of driving, more honks than the Tower of Babel—"Let me out of here!" The man jabbered angrily, talking about how long it takes to fix a dented bumper, but I couldn't think of any response. "One hundred," I said again.

A crowd gathered, and the cabbie began to play to the jury—it was a bad dent; he worked long days; repairs took time. And then a tiny old woman stepped forward and touched his arm. "Take the money," she said softly. The cabbie looked down at her—she couldn't have been more than five feet tall—and fell silent. He didn't say a word when I handed him the bill.

In the Capital Motors lot, Mr. Wang ran a finger along the dent. "No problem!" he said.

"Look, I'm happy to pay for it," I said.

"You're an old customer," he said. "Forget it." We shook hands and I left him at the front desk, smoking a cigarette beneath the eternal sign:

CUSTOMER SATISFACTION RATING: 90%
EFFICIENCY RATING: 97%
APPROPRIATE SERVICE DICTION RATING: 98%
SERVICE ATTITUDE RATING: 99%

# HOME AND AWAY

LITTLE FATTY KEPT LEAVING IT SHORT. TWICE HE DROPPED THE basketball on the way up, and the third time, when Yao Ming finally lifted him above the rim, he held the ball too low. His name was Sun Haoxuan; he was four years old; he weighed fifty-nine pounds; and he had been selected by an advertising firm that had recently scouted Beijing kindergartens for a fat boy with round cheeks and big dark eyes. There was a substantial talent pool. In Chinese cities, rising standards of living have combined with the planned-birth policy in a way that recalls the law of conservation of mass: there are fewer children, but often there is more child. It's common for adults to refer to these kids as simply Xiao Pangzi—Little Fatty. "Get Little Fatty ready!" the director shouted whenever he needed Sun Haoxuan. "Move Little Fatty back two steps!"

We were at the Beijing Film Studios, where Yao Ming was shooting a television commercial for China Unicom, a telecommunications company. The script was simple: fat child meets seven-foot-six-inch basketball player; basketball player lifts fat child; fat child dunks. What had not been factored in was Little Fatty's behavior. He squirmed away at every opportunity; sometimes he pointed directly at Yao Ming and announced, with an air of sudden revelation, "Yao Ming!" For half an hour, the adults in the studio—cameramen, assistants, tech guys—had been silently aiming ill wishes his way, and maybe that was why, on the fourth take, Yao stumbled and accidentally rammed Little Fatty's nose against the rim.

The sounds came in quick succession: a soft thud, a dropped ball—*bounce, bounce, bounce-bounce*—and then the child began to wail.

Little Fatty's mother rushed over, and Yao Ming stood helplessly, shoulders slumped. He was breathing hard. Somebody wiped Little Fatty's face—no blood, no foul. On the next take, he finally dunked the ball, and there was a thin round of applause. Yao wandered over to the edge of the set, where I was standing, and said, in English, "Weight training."

He had just finished a sensational rookie season for the Houston Rockets. Now, in the summer, the twenty-two-year-old center had returned to China with one clear objective: to lead the national team to the title in the Asian Basketball Championships, which serves as the regional qualifier for the Olympics. Usually, China dominates Asian basketball, but there were serious challenges this year. Wang Zhizhi, the country's second-best player, had not come back from America because of political problems. Yao Ming had become involved in a high-profile lawsuit, which was interpreted by the Chinese press as a clash between the rights of the individual and the state. Increasingly, Yao's world was divided: there was the sanctity of the sport, and then, off-court, a whirlwind of distractions, ranging from the burdensome to the bizarre. When I had last visited him, in July, he was staying with the Chinese team in a hotel in Qinhuangdao, a seaside town that was hosting an exhibition game against a squad from the United States Basketball Academy. Yao didn't play—he had just received eight stitches in the eyebrow after a teammate elbowed him in practice. Before the game, a China Unicom representative with a digital recorder coached Yao through a series of phrases that would be sold as alarm messages to mobile-phone subscribers. "Wake up, lazy insect!" Yao said obediently, and then his bandaged brow dipped when the woman asked him to repeat it ("More emphasis!").

That evening, the Chinese nearly threw the game away—in the final quarter, they couldn't handle a full-court press from the ragtag American team. "I think the center needs to come to half court against the press," Yao told me afterward, in his hotel room. Liu Wei, the Chinese point guard and Yao's best friend, was sprawled on one bed. Yao sat on the other bed, which had been crudely extended: the head consisted of a wooden cabinet covered by blankets. We spoke in English; he talked about the NBA off-season news that he had culled from the Internet. He had not spoken to any of his Houston Rocket teammates since returning to China. "Did you hear about Rodman?" Yao said. "He might come back. I can't believe the Lakers got Payton and Malone. I can't believe they only spent four million. If Kobe is OK, it's like a Dream Team." The names sounded foreign and faraway—Mark Cuban, Shaq, Kirilenko. "AK-47," Yao said, using the sports-talk nickname for Andrei Kirilenko, a Russian forward on the Utah Jazz. Yao smiled like a kid at the sound of the phrase. "AK-47," he said again.

Yao Ming weighed ten pounds at birth. His mother, Fang Fengdi, is over six-two; his father, Yao Zhiyuan, is six-ten. Both were centers: he played for the Shanghai city team, and she was on China's national team. Chinese sports couples aren't uncommon—Yao Ming is dating Ye Li, a six-two forward on the women's national team. When Yao was growing up, the apartment directly overhead was home to the Sha family, whose parents had both been point guards for Shanghai teams. "My mother and father were introduced by the basketball organization," Sha Yifeng, a childhood friend of Yao Ming, told me. "In the old days, that's how they took care of your life."

Today, Yao's parents are in their early fifties, trim and black-haired, and they carry themselves with the physical dignity of former athletes. But they speak about basketball with distinct detachment. Neither played the game as a child; sports were a low priority for China in the 1960s, particularly during the early years of the Cultural Revolution. Later, officials began to restore the national sports system, scouting for height to fill out the basketball rosters. Yao Zhiyuan began to play at the age of nineteen. Fang Fengdi was discovered at sixteen. "To be honest, I didn't much like it," she told me, when I met them both in Shanghai. "I wanted to be a dancer or an actress." By 1970, she was traveling to games around the world with the national team. "I didn't think of it as something I did or didn't want to do," she told me. "I thought of it as a responsibility. It was a job."

In China, competitive sport is a foreign import. Traditional physical activities like wushu and qigong are as much aesthetic and spiritual as they are athletic, and Chinese historians say that "modern sport" began with the 1839–1842 Opium War. In the following decades, as foreign traders and missionaries established themselves in treaty ports, their schools and charitable institutions introduced Western competitive sports. American missionaries brought basketball to China at the end of the nineteenth century. At the same time, the Chinese were struggling to overcome foreign occupation, and soon they began to see sports as a symbolic way to avenge the injustices of the past century. The goal was to beat the foreigner at his own game. After the Communists came to power, in 1949, they established a state-funded sports-training system modeled on the Soviet Union's. Promising young athletes were recruited for special "sports schools."

In first grade, Yao Ming was taller than his teacher. By the time he entered third grade, he was five-seven, and Shanghai's Xuhui District Sports School selected him for its after-school basketball program. Recently, I visited Yao's first coach, Li Zhangming, who, like a traditional Chinese educator, spoke of his former player in completely unsentimental terms. ("He didn't much like basketball. He was tall, but slow and uncoordinated.") After our conversation, I wandered around the basketball courts of Shanghai's No. 54 Middle School, where the Xuhui Sports School holds some of its practices. I watched a group of young girls performing basketball drills; after a while, I introduced myself to the coach, a tall woman named Tao Yanping.

"I was a teammate of Yao's mother," Tao said. "I went to their wedding. I remember giving them towels and thermoses—things you gave newlyweds back then. See that girl there?"—a red-faced child, the tallest on the court—"Her mother was also my teammate. That girl is in the third grade. Her mother is 1.83 meters tall, and she made the national team."

I asked Tao how she recruited. "We go to the schools and look at the children's height, and then we check their parents' height."

The two-hour practice consisted mostly of ball-handling drills. Tao was attentive, shouting commands at her charges. ("Little Swallow, you're traveling! Who taught you to do that?") At the end of the practice, tall parents materialized at courtside. "I only want her to play because it's good for her health," Zhang Jianrong, a woman who was nearly six feet tall, told me. She explained that basketball was a good activity for her daughter after a day in school, but homework was more important. Like the other parents I met, Zhang was middle-class, and none of them expressed a wish for their child to have a future in sports. They were basketball moms in a country that selects its basketball moms by height—China cannot yet afford to provide every public school with coaches and sports facilities.

Instead, the key to the nation's sports strategy has been early recruitment and focused training on a relatively small group of potential athletes. This system has proved effective in low-participation, routine-based sports like gymnastics and diving, but when it comes to basketball it may be China's greatest weakness. In America, where community leagues are common and school coaches are plentiful, athletes emerge from an enormous pyramid of participants. Some, like Allen Iverson, rise to the top with remarkable passion and creativity—but if a recruiter had shown up at the Iverson home when Allen was in the third grade, he would have found no father and a short mother who had given birth at the age of fifteen. It's significant that China has yet to produce a great male guard—the position requires skill and intensity rather than height. All three Chinese players currently in the NBA are centers, and two are second-generation centers. The Chinese national team is notorious for choking in key games, partly because the ball handling is inconsistent. Players rarely appear to enjoy themselves, and their character has not been formed by true competition; even as free-market reforms have changed many Chinese industries, the sports world is a throwback to socialism, with its careful planning and career stability. Once, when I asked Yao Ming how many Chinese would be in the NBA in a decade, he said only three or four.

Throughout Yao Ming's childhood, his parents emphasized that basketball was a hobby, not a career. As a boy, he liked history, geography, and archaeology. "When I was small, I always wanted to be famous," Yao once told me. "I thought I'd be a scientist or maybe a political figure. It didn't matter, as long as I was famous." In sixth grade, he grew taller than his mother. He surpassed his father's height in ninth grade. By then, he was already under contract to the Shanghai Sharks youth team. When he was seventeen, and seven-two, Yao Ming joined the Chinese national team. Relatives told me that it wasn't until then that his parents resigned themselves to his career as a professional athlete.

Once, I asked Fang Fengdi if there had been a moment when she first sensed that basketball inspired Yao Ming. It was the only time she smiled when discussing the sport. I sensed that she was talking about herself—the woman who once wanted to be a dancer or an actress—as much as she was talking about her son. She said, "The Harlem Globetrotters came to Shanghai when he was in elementary school. Tickets were really hard to get—I was able to find only two. That wasn't just basketball for a competition or a job. I remember thinking, Americans are good at enjoying themselves! Those players took a normal sport and turned it into something else—a performance. Afterward, I could tell that Yao Ming was inspired. It made a deep impression on him."

THE FIRST MALE PLAYER TO MAKE THE JUMP FROM MAINLAND China to top-level American basketball was Ma Jian, a forward who played at the University of Utah for two years in the 1990s. Ma noticed that during Utah's pre-game meetings, an assistant coach sometimes wrote a *W* or a *B* on the chalkboard next to an opposing player's name. "The white players were shooters," Ma explained to me, when we met recently in Beijing. "If he put a *B* there, we knew they were athletes." Ma never saw a *C* on the board. In 1995, he tried out for the Los Angeles Clippers. "The first time I stepped onto the team plane in the pre-season, I saw the blacks sitting on one side and the whites on the other. I looked at myself—should I go on the brothers' side or the whites' side? Finally I said, Just play."

In 2002, after the Rockets selected Yao Ming with the first pick in the NBA draft, it was less than a week before somebody in the league made a remark that could be construed as racist. During a television interview, Shaquille O'Neal, the NBA's dominant center, announced: "Tell Yao Ming, 'Ching chong yang wah ah so.'" O'Neal's joke went largely unnoticed at the time, but it was resurrected in January the following year, when a columnist for *Asian Week* attacked O'Neal for it.

The column sparked a media frenzy shortly before Shaq and Yao's first on-court meeting. But Yao immediately defused the controversy. "There are a lot of difficulties in the two different cultures understanding each other," he said. "Chinese is hard to learn. I had trouble with it when I was little." The NBA released a statement pointing out that the league included players from thirty-four countries. By game time, the issue was all but dead. The Rockets won by four points, in overtime; O'Neal outplayed Yao, but the Chinese center had a spectacular start and held his own. Afterward, O'Neal told the press, "Yao Ming is my brother. The Asian people are my brothers."

That season, I spent most of a month following Yao's games, and people repeatedly brought up the O'Neal incident. None of the black fans I talked to had anything bad to say about Yao—many believed that he brought something fresh to American sports. "It's not like normal, where people say, well, he's a black athlete, so he moves like this, or he's a white athlete, so he shoots like that," Darice Hooper, a physical therapist who was attending the All-Star Game in Atlanta, told me. "It's like we had an alternative."

Juaquin Hawkins, one of Yao's teammates on the Rockets, agreed. "It's not just people thinking, I'm rooting for him because he's African-American, or I'm rooting for him because he's white," Hawkins told me. "They're rooting for him as a person."

Hawkins was familiar with the outsider's role. A native of Lynwood, California, he had failed to make the NBA in 1997, and the following year he wound up playing professionally in Chongqing, deep in the Chinese interior. I had lived in the same region, and Hawkins laughed when I mentioned the basketball slang there. If a player shoots an air ball, the fans shout "*yang-wei*"; in the Sichuanese dialect, it means "impotent." To encourage the home team, they chant "*xiongqi*"—"erection."

There are few foreigners in Chongqing, and even fewer blacks. I asked Hawkins how he had coped with being so different. "I always felt like I was representing my heritage," he said. "Lynwood is next to Compton—it's basically like coming from Compton. There's a lot of negative things said about that area, and that's something I take with me wherever I go. But I had a good childhood. I was raised by my mother. I try to represent that."

An uncle had introduced Hawkins to basketball as a child; he had never met his father. "All I know is his first name, and the fact that he didn't want to deal with having a family," Hawkins said. "It's a sad story, but I've used that as motivation." He met his wife through basketball—both had played at Lynwood High School, and then at Long Beach State.

Hawkins said that Chongqing was the roughest place he had ever lived. He had also played professional basketball in Taiwan, Japan, and the Philippines. He had toured with the Harlem Globetrotters. ("That was actually real beneficial.") Twice, he had attended NBA pre-season camps, only to be cut. In the summer of 2002, hoping to have one last chance to attract NBA interest, Hawkins purchased a two-headed VCR, which he used to create a highlight reel of clips from all the places he had played. The Rockets invited him to camp, where he established himself as a defensive specialist and beat out two other unsigned players for a roster spot. At twenty-nine, he was the oldest rookie in the league to make an opening-day lineup. When Hawkins learned that he was on the team, he telephoned his mother and wept.

SUCCESSFUL ATHLETES ARE INEVITABLY DISPLACED—IF YOU'RE good, you leave home—and something is always lost in transition. Much of what Hawkins carried onto the court would have been invisible to Chongqing fans, who know nothing about Compton or African-American single-parent families. In Chongqing, Hawkins was simply an excellent player who looked completely different from everybody else in the city. When I lived in a nearby town, it was common for crowds of twenty or more people to gather and gawk at me on the street. A local nightclub once hired an African dancer, knowing that his freakishness would draw customers.

Yao Ming had an excellent rookie season, and there were signs that eventually he would develop into a dominant center. But the Rockets ran only about thirty plays a game to him; initially, his American fame resulted from his height and his off-court persona. He handled attention with remarkable humor and grace, and he provided an escape from the undercurrent of racial tension in American sports. He also appealed to the national missionary instinct: if Americans had failed to convert the Chinese to God and democracy, at least they were turning them into NBA fans. The American media portrayed him as a nonthreatening figure—a gentle giant.

But he entered another world whenever he dealt with the Chinese press. After a difficult defeat in Los Angeles, where Yao had fouled out for the first time in his NBA career, a Chinese reporter asked what it had been like to be dunked on by Kobe Bryant. Yao said evenly, "Please don't ask me about an incident in which I have no face." At an All-Star Game press conference, Yao showed up wearing an old Chinese national-team sweatshirt, and a Chinese reporter asked why. "It's comfortable, that's all," Yao said. Another reporter asked, "If you could say one sentence to all of the young Chinese players back home, what would you say?" Yao's sentence: "I don't believe that I can say very much with one sentence."

Even as they idolized him, few people in China seemed to realize how different Yao was from the typical Chinese athlete. When he played, the joy was apparent on his face. He hit free throws in the clutch, and the Rockets learned to run plays to him at the end of close games. Often, he subtly deflected the patriotic questions of the Chinese media, as if sensing that such concerns were too heavy to bear on the court.

The Chinese motivation for sport is so specific and limited—the nationalism, the sports schools—that it rarely survives a transplant overseas. Athletics has meant little to most Chinese-American communities, including the one in Houston, which has grown rapidly in the past decade. The city has an estimated fifty thousand Chinese, as well as large numbers of ethnic Chinese from Vietnam. Houston's Chinese tend to be highly educated, with an average annual household income over fifty thousand dollars a year, higher than the city's average.

The largest Asian district in Houston is along Bellaire Boulevard—a six-mile strip-mall Chinatown. In February, I spent two afternoons driving along Bellaire, where some of the signs reminded me that locals were adjusting to a new culture (All Stars Defensive Driving); others reflected success (Charles Schwab, in Chinese characters); and some were distinctly Chinese (a lot of beauty parlors—the Chinese are meticulous about hair). But I couldn't find a basketball. Though everybody loved Yao Ming, people told me that the children in the community didn't play sports much; they were too busy studying. I searched for hours before finding a sporting-goods store— Sports Net International, in a mall called Dynasty Plaza—and they stocked gear only for racquet sports. "The Chinese are not so interested in basketball, because of their size," David Chang, the owner, told me. "But if you're interested in Yao Ming, you should talk to the people at Anna Beauty Design. They cut his hair."

It wasn't exactly what I'd been searching for, but I figured I'd see what they had to say. A Taiwanese woman sat behind the receptionist's desk. I asked if Yao Ming got his hair cut there.

She paused before answering. "No," she said. "Yao Ming does not get his hair cut here."

I tried again. "Does somebody from Anna's go to Yao Ming's home to cut his hair?"

"That's something I can't answer," she said coyly. A moment later, the manager walked in. "This guy's a reporter," she told him. "He wants to know if we cut Yao Ming's hair."

The manager shot me a dirty look. "Don't tell him we do that," he said.

The receptionist added, exactly five seconds too late, "He speaks Chinese."

All told, I tracked down three defensive-driving schools, three bookstores, six banks, and fourteen beauty salons—but no *lanqiu*. In Houston's Chinatown, it was easier to find Yao Ming's barber than a basketball.

AT THE END OF FEBRUARY, THE ROCKETS EMBARKED ON AN important East Coast road trip. Their final game of the trip was against the Washington Wizards; both teams were fighting to make the playoffs in their respective conferences, and Yao Ming was in the running to be named Rookie of the Year. This would be the final meeting between Yao Ming and Michael Jordan, who was retiring in order to return to his position as president of the Wizards.

The night before the Washington game, the Chinese embassy hosted a special reception for Yao. It was a snowy evening, and I caught a cab to the embassy. The driver, a seventy-five-year-old black man named Willard Cooper, asked why I was going there. "I can really relate to how the Chinese people feel about him," Cooper said, when I mentioned Yao. "That's the way I felt years ago, when Jackie Robinson was playing."

At the embassy, Chinese food and Yanjing beer were served—the Beijing-based brewery had signed a Rockets sponsorship after Yao Ming was drafted. The big meeting room was packed with people: diplomats and émigrés, Sinophiles and market analysts. Scraps of conversation floated in the air.

"Yanjing paid six million dollars. Their distributor is Harbrew."

"Who gives a sixty-year distribution contract? But you know, from the Chinese point of view, it's a stream of production. They don't understand the concept of branding."

"He's been in China fifteen years as a value-added player."

"Actually, I'm with the White House press office."

"You know, Anheuser-Busch owns 27 percent of Tsingtao."

"There he is! Did you get a picture?"

"Imagine being that tall!"

A round of applause followed Yao into the room. Lan Lijun, the minister of the embassy, gave a short speech. He mentioned Ping-Pong diplomacy and "the unique role sports have played in bringing our countries together." In closing, he said, "We have full confidence that China and the United States will work together to continue to improve our bilateral relations."

Yao, in a gray suit, stooped to reach the microphone. Behind him, a display case held a ceramic horse from the Tang dynasty. Red lanterns hung from the ceiling. Yao spoke for less than a minute, and he didn't say anything about Sino-American relations. "Seeing all these lanterns reminds me of home," he said softly. "Growing up, my impression of the Chinese embassy was like a fantasy, something you see on television and in the movies." There was a rush for autographs, and staff hustled Yao into a back room. In the corner, a pretty Eurasian girl in a red dress was crying. Her parents said that Yao had walked past without signing her invitation. "He's her favorite player," the mother told me, adding that the girl had been adopted from Uzbekistan. A staff member took her invitation, promising to get a signature.

Yao was at the embassy for nearly two hours. After he left, people lingered in groups, chatting and drinking Yanjing. We had reached the Sino-American witching hour—the Chinese guests, always prompt, were gone, but the Americans lingered in the way that Americans do. I found myself standing next to Chen Xiaogong, the defense attaché. Chen was glassy-eyed; he kept touching his watch. "I'm surprised so many Americans know Yao Ming," he murmured.

THE NEXT NIGHT, KHA VO SINGS FRANCIS SCOTT KEY AND Michael Jordan comes out hot. Four baskets in the first quarter: turnaround, jump shot, jump shot, turnaround. Ten days ago,

Jordan celebrated his fortieth birthday, and since then he's been averaging nearly thirty points a game. Yao works against Brendan Haywood, the Wizards' seven-foot center. Haywood looks short tonight. Six points for Yao in the first quarter; Rockets down by nine. Sold-out arena: twenty thousand plus. Lots of Asians—red flags in the upper levels.

Second quarter: Rudy Tomjanovich, the Rockets coach, plays a hunch and goes with Juaquin Hawkins, who rarely sees action. Hawkins nails a twenty-footer, then a three-pointer. He draws a charge and steals a pass. Hawkins looks hungry, as if he'd just escaped from Chongqing: he hasn't scored in nine days. Moochie Norris runs the point for the Rockets. Moochie has cornrows, a barrel chest, and four Chinese characters tattooed on his left wrist: *huan de huan shi*. ("Never satisfied," he once told me, when I asked what it meant, and then I crossed to the other side of the locker room. "It actually doesn't have a very good meaning," Yao said, in Chinese. "Basically, you'll do whatever it takes to protect yourself.") Yao doesn't score in the second quarter. Jordan has eighteen. Rockets down by twenty. Halftime show: Chinese lion dance, followed by an announcement about Black History Month.

Houston sleepwalks through the third. At one point, they trail by twenty-four. In the final quarter, Maurice Taylor, a Rockets forward, starts to hit jumpers. With six minutes to go, Houston down by fourteen, Tomjanovich brings in Yao, and the game turns. Hawkins sinks a three, then knocks the ball loose from Tyronn Lue. The two players collide and Lue falls, writhing in pain. Separated shoulder, cut eye: good night, Tyronn. Four straight baskets by the Rockets. In the final three minutes, Yao steps to the free-throw line four times, and makes everything. Haywood fouls out. Overtime.

Hawkins guards Jordan, and they trade baskets to start the extra period. Yao hits a baby hook to give the Rockets the lead. The Wizards feed Jordan every time down the court, and now, after playing for forty-five minutes, he suddenly finds new life. Turnaround jumper over Hawkins. Next possession: Jordan crossover dribble to his left; Hawkins freezes—dunk. Next possession: Jordan hard drive; Hawkins falls, no call—jumper. Next possession: Jordan drives; Hawkins lags, Yao goes for the block—goaltending. Jordan scores ten in overtime and finishes with thirty-five points and eleven rebounds. Yao has sixteen and eleven; Hawkins scores ten. In the final seconds, with the Rockets down by two, Yao gets a defensive rebound and, instead of calling a time-out, throws the outlet pass. Bad shot. Rockets lose.

AFTER THE GAME, IN THE ROCKETS' LOCKER ROOM, HAWKINS sat alone on a bench. "It was frustrating," he told me. "He's the greatest player ever." Outside, Tomjanovich had given his guard credit for keeping Houston in the game. "Hawkins came in and gave us life defensively," he said. "Hawkins and Mo Taylor were the guys today."

Yao sat in front of his locker, a towel wrapped around his waist. The Chinese media pressed close, and he told them that he should have called the time-out.

In the Wizards' locker room, I joined a group of reporters waiting for Jordan. Unlike the other players, he never met the press while showering and changing, and the team always arranged a special podium for their star. Jordan stepped behind the microphone, dressed in a gray pin-striped suit. Somebody asked if the Wizards would make the playoffs. "I've never had a doubt that we would," Jordan said.

Another reporter asked about the overtime period, and Jordan was dismissive of Hawkins: "I was going against a young kid who didn't really know how to play, and he tried a couple of flops."

Somebody asked about Yao. "You can sit here and talk about how good he eventually could be," Jordan said. "But at some point, he's going to have to showcase what everybody expects. He's going to have to get better, which I think he will."

Kobe Bryant was having a spectacular season, and a reporter asked how Jordan in his prime would have fared against the Lakers' guard. Jordan praised Bryant, but then he grinned: "I think I would have a good chance of taking care of business."

Jordan spoke bluntly, and he saw the game with athlete's myopia: on the court, it didn't matter where the players had come from or where they were going. For fifty-three minutes, the competition was more important than everything else that surrounded it. But like most games, it soon receded into the essence of statistics—the meaningless points, the pointless minutes. In the end, neither the Wizards nor the Rockets made the playoffs. Michael Jordan never again collected thirty points and ten rebounds in a game, and in May, after retiring, he was forced out of the Wizards organization. Less than three weeks after the Washington game, Rudy Tomjanovich was diagnosed with bladder cancer, and he stepped down as coach. Yao Ming did not win Rookie of the Year. And the following season Juaquin Hawkins, after failing to make an NBA team, rejoined the Harlem Globetrotters.

ALTHOUGH IT IS DIFFICULT FOR A CHINESE ATHLETE TO COME to America, it may be even harder for him to return home. Ma Jian, who tried out for the Los Angeles Clippers without the explicit blessing of Chinese authorities, was never allowed back onto the national team. Wang Zhizhi, a seven-one center who dominated Chinese basketball in the late 1990s, had even more trouble. While Wang was rising as a player, the Communist Party was restructuring many of its sports bureaus into for-profit entities. The Chinese Basketball Association hoped to become self-sufficient through corporate sponsorships and income from its professional league, known as the CBA. In this climate, the CBA became a strange beast: its sponsors included private companies, state-owned enterprises, and the People's Liberation Army, which ran a team called the Bayi Rockets. Wang Zhizhi played for Bayi, and in 1999 the Dallas Mavericks selected him in the second round of the NBA draft. For nearly two years, Dallas courted Wang's bosses, trying to convince them to let the player go. Wang was officially a regimental commander in the PLA.

In the spring of 2001, Dallas and Bayi finally came to an agreement, and Wang became the first Chinese to play in the NBA. He was twenty-three years old. In the off-season, he returned home, as promised, representing both the national team and Bayi. But after Wang's second NBA season, in which he averaged about five points a game, he requested permission to delay his return to China so that he could play in the NBA's summer league. He promised to join the national team in time for the World Championships in August.

The Chinese national team is notorious for its grueling practice schedule—twice a day, six days a week. Fear shapes the routine: coaches know that they will be blamed if the squad loses without logging massive hours. Any innovation is resisted. Before games, the Chinese men's team warms up by conducting the same rudimentary ball-handling drills that I watched the third-grade girls perform in Shanghai.

In the summer of 2002, Chinese authorities refused Wang's request, but he stayed in the United States anyway. Dallas did not offer him a contract, reportedly in part because they did not want to ruin the good relationship that they had developed with the Chinese. In October, Wang signed a three-year, $6 million contract with the Los Angeles Clippers. After that, Clippers games were banned from Chinese television (NBA broadcasts often draw more than ten million viewers in China). The ban turned Wang into a marketing liability—one NBA general manager told me that teams were wary of signing him in the future.

Wang, whose military passport had expired, reportedly re-

ceived a U.S. green card. Over the summer, he tried to negotiate a return to China, asking for a new civilian passport and a
guarantee that he could come back to the NBA after the Asian
Championship. The chain of communication had grown so
complicated that Wang relied heavily on a Chinese sportswriter
named Su Qun to contact PLA leaders and basketball officials.
"I know that as a journalist I should stay out of this," Su, who
writes for Beijing's *Titan Sports Daily*, told me. "But I happen to
be close to Wang. We have to save him, like saving Private Ryan."

Wang, who declined my request for an interview, did
not return to China. I spoke about him with Li Yuanwei, the
secretary-general of the Chinese Basketball Association. "Wang
has placed too much emphasis on his personal benefit," Li told
me. "I assured him that there is no risk. The PLA also assured
him. But he doesn't believe us, and he keeps demanding conditions that are not necessary. It's very sad."

Wang's problems formed a troubling backdrop to Yao
Ming's move to the NBA. Before leaving China, Yao promised
to fulfill his national-team commitments during the off-season,
and he reportedly agreed to pay the Chinese Basketball Association 5 to 8 percent of his NBA salary for his entire career. He
also paid the Shanghai Sharks, his CBA team, a buyout that was
estimated to be between $8 million and $15 million, depending
on his endorsements and the length of his career. Yao's four-year
contract with the Rockets was worth $17.8 million dollars, and
after one season his endorsement income was already higher
than his salary.

But even Yao's sponsorship potential was threatened by the
irregularities of China's sports industry. In May, Coca-Cola issued a special can decorated with the images of three national-
team players, including Yao, who already had a contract with
Pepsi. The basketball association had sold Yao's image to Coca-
Cola without the athlete's permission, taking advantage of an
obscure sports-commission regulation that grants the state the

right to all "intangible assets" of a national-team player. The regulation appeared to be in direct conflict with Chinese civil law. Yao filed suit against Coca-Cola in Shanghai, demanding a public apology and one yuan—about twelve cents. The Chinese press interpreted the lawsuit as a direct challenge to the nation's traditional control of athletes.

When I spoke with Li Yuanwei, of the basketball association, he emphasized that Coca-Cola was an important source of funding, and he hoped that the company and Yao would reach an agreement out of court. Li told me that Americans have difficulty understanding the duties of an athlete in China, where the state provides support from childhood. I asked if the same logic could be applied to a public school student who attends Peking University, starts a business, and becomes a millionaire. "It's not the same," Li said. "Being an athlete is a kind of mission. They have an enormous impact on the ideas of the common people and children. That's their responsibility."

Before I traveled to Harbin, in northeastern China, to attend the Asian Championship, I talked with Yang Lixin, a law professor at People's University in Beijing. Yang was preparing a seminar on the Coca-Cola case. "Contact with American society probably gave Yao some new ideas," Yang told me. "It's like Deng Xiaoping said—some people will get wealthy first. Development isn't equal, and in a sense rights also aren't equal. Of course, they are equal under law, but one person might demand his rights while another does not. It's a choice. In this sense, Yao Ming is a pioneer."

DISPLACED PEOPLE HAVE ALWAYS WANDERED TO HARBIN. DURING the twentieth century, they came and went: White Russians, Japanese militants, the Soviet Army. Even today, much of the architecture is Russian. Harbin's symbol is the former St. Sofia Church: gold crosses, green onion domes, yellow halos around white saints. The city has one of the last Stalin Parks in China.

At the end of September in 2003, sixteen teams arrived for the Asian Championship; the winner would qualify for the Olympics. The squads defied any simple concept of nationality and border. Most of the Kazakhstan players were in fact Russians whose families had stayed after the collapse of the Soviet Union. The Malaysian team had a peninsular range: ethnic Chinese, Indians, Malays. Qatar included athletes from Africa and Canada—opponents grumbled that they had loosened the definition of a Qatari. The Syrian coach was a black man from Missouri; the Qatar coach was a white man from Louisiana. Iran's coach was a Serb who told me that his playing career had been cut short; he pulled up his sleeve to reveal a cruel scar. ("Not long after that, I started coaching.")

Except for the Chinese team, everybody stayed at the Singapore Hotel. Tall people in sweat suits lounged in the lobby; a restaurant on the second floor had been converted into a halal cafeteria. The South Korean team included Ha Seung-Jin, an eighteen-year-old who was seven-three, weighed 316 pounds, and had basketball bloodlines—his father was once a center for the Korean national team. People expected Ha to be a first-round NBA draft pick the following year, which would make him the first Korean to play in the league. "I want to be a Korean Yao Ming," he told me, through an interpreter (who added that the young player's nickname was Ha-quille O'Neal). Ha was eager to play Yao; everybody expected China and South Korea to meet in the finals. Last year, in the Asian Games, South Korea had upset the Chinese. Ha hoped to get Yao into foul trouble. "Yao Ming likes to spin to his right," Ha said. "I'll establish position there and draw the foul."

The other seven-three player in the tournament was an Iranian named Jaber Rouzbahani Darrehsari. Darrehsari had played for only three years, since being discovered in the city of Isfahan, where his father sold fruit and vegetables in a market. Darrehsari's wingspan was over eight feet wide. Once, when he was leaving the court after a game, I asked him to touch the rim. He hopped ever so lightly, and then stood still: fingers curled around the metal, the balls of both feet planted firmly on the hardwood. He was seventeen years old. He had dark, long-lashed eyes, and he hadn't yet started shaving—it was as if a child's head had been attached to an elongated body with dangling arms. In Iran's first two games, Darrehsari played only a few minutes; smaller opponents shoved him mercilessly. He looked terrified on the court. Sitting on the bench he almost never smiled.

THE CHINESE TEAM STAYED AT THE GARDEN HAMLET HOTEL, a walled compound reserved for central-government leaders. All summer, Yao had been unable to appear in public without attracting a mob, and in August, the Chinese media reported that a medical exam had revealed that Yao had high blood pressure. His agents said the condition was temporary, but there was concern that all the pressure and the excessive practice schedule could shorten Yao's career. He issued a message on his website, expressing frustration with the national team: "I have been exhausted because of the poor security at the national team games . . . too many public appearances and commitments by the Chinese national team, and incessant fan disturbances at the team hotel."

A few hours before China played Iran, one of Yao's agents told me that I could meet with his client. Yao was represented by an entity known as Team Yao, which consisted of three Americans, two Chinese, and one Chinese-American. Half the team had come to Harbin—Erik Zhang, Yao's distant cousin and the team leader; John Huizinga, a deputy dean at the University of Chicago Graduate School of Business, where Zhang was a student; and Bill A. Duffy, who headed BDA Sports Management. They were accompanied by Ric Bucher, a senior writer for *ESPN the Magazine*, who had signed on to write the official Yao biography. A day earlier, Yao had agreed to a multiyear endorsement contract with Reebok, which had yet to be announced. A source close to the negotiations told me that the deal, which was heavy with incentives, could be worth well over $100 million—potentially, the largest shoe contract ever given to an athlete.

A security guard let me into the compound. I walked through rows of willows, past well-kept lawns decorated with concrete deer. It was raining hard. Despite being as much as $100 million richer than he was the day before, Yao still did not have a bed that fit. This time, the hotel had arranged for a wooden cabinet at the foot of the bed, to prop up his legs. The shades were drawn; discarded clothes lay everywhere. Liu Wei, the point guard, sprawled in a tangle of sheets on the other bed.

The night before, after China had defeated the Taiwanese team by sixty-one points, Yao had sprained his left ankle while boarding the team bus. Now Duffy, a former player in his forties, was examining him. The ankle was slightly swollen. Duffy told Yao to ice it immediately after that night's game. Yao answered that there was no ice at the arena.

Duffy looked up, incredulous. "They don't have ice?"

The games were being held in a converted skating rink, in a multi-sports complex, less than two hundred miles from the Siberian border.

"No ice," Yao repeated, and then he spoke in Chinese to Zhang: "I've been getting acupuncture."

After a few minutes, Team Yao left the room. Yao and I chatted in Chinese about the tournament, and then I mentioned that his first coach had told me that Yao didn't like basketball as a child. "That's true," Yao said. "I didn't really like it until I was eighteen or nineteen."

I asked Yao about his first trip to the United States, in 1998, when Nike had organized a summer of training and basketball camps for him. "Before then, I was always playing with people who were two or three years older than me," he said. "They were always more developed, and I didn't think that I was any good. But in America I finally played against people my own age, and I realized that I was actually very good. That gave me a lot of confidence."

He talked about how difficult it had been when he first moved to Houston ("Everything about the environment was strange"), and I asked him about the differences between sport in China and America.

"In China, the goal has always been to glorify the country," Yao said. "I'm not opposed to that. But I personally don't believe that that should be the entire purpose of athletics. I also have personal reasons for playing. We shouldn't entirely get rid of the nationalism, but I do think that the meaning of sport needs to change. I want people in China to know that part of why I play basketball is simply personal. In the eyes of Americans, if I fail, then I fail. It's just me. But for the Chinese, if I fail, then that means that thousands of other people fail along with me. They feel as if I'm representing them."

I asked about the pressure. "It's like a sword," he said. "You can hold it with the blade out, or with the blade pointing toward yourself." Then I mentioned Wang Zhizhi's situation.

"There's an aspect of it that I shouldn't talk about," Yao said slowly. "It's best if I simply speak about basketball. If Wang were here, it would be good for me. I just know that if he played, I wouldn't feel as if so much of the pressure was falling onto one person."

I asked about the Coca-Cola lawsuit. "I always put the nation's benefit first and my own personal benefit second," Yao said. "But I won't simply forget my own interests. In this instance, I think that the lawsuit is good for my interests, and it's also good for other athletes. If this sort of situation comes up in the future for another athlete, I don't want people to say: 'Well, Yao Ming didn't sue, so why should you?'"

* * *

No pre-game national anthems at the Asian Championship. Before tonight's game, the loudspeaker plays an instrumental version of the theme from *Titanic*. The Iranians look nervous. Sold-out arena: four thousand plus. The crowd is full of thunder sticks—they are, after all, manufactured in China—but nobody seems to know how to use them. The lack of noise feels like intense concentration. The spectators cheer both sides—enthusiastically when the Chinese score, politely for an Iranian basket.

The coach plays a hunch and starts Darrehsari, who looks scared. On every possession, the Iranians avoid Yao's lane, swinging the ball along the perimeter: Eslamieh to Bahrami to Mashhady. Mashhady to Bahrami to Eslamieh. Yao does not score for nearly six minutes. At last, he brushes Darrehsari aside, grabs an offensive board, and dunks with both hands. Tie game. Next possession: China lead. Next possession: bigger lead. Eslamieh to Bahrami to Mashhady. Somebody throws it to Darrehsari, fifteen feet out. Yao doesn't bother to challenge. The shot develops as a chain reaction across the entire length of Darrehsari's seven-three frame: knees bend, waist drops, elbows buckle, long hands snap—*swish*. Running back down the floor, he tries to fight back a smile. A few possessions later, he fouls Yao hard. Darrehsari is all elbows and knees, but for the first time in the tournament he looks like he wants to be on the court. The coach plays him the entire half. He scores four and leads Iran with four rebounds. After the halftime buzzer, his teammates clap him on the back.

Yao plays half the game: fifteen points, ten rebounds. He looks bored. China wins by twenty-four. Later, Yao tells me diplomatically that Darrehsari has potential. "It depends on environment," Yao says. "Coaching, teammates, training." For the rest of the tournament, Darrehsari does not play half as many minutes. The day after the China game, he beams and tells me, "It was an honor to play against Yao Ming."

BEFORE THE FINAL, CHINA UNICOM UNVEILED ITS NEW COM-mercial at a press conference attended by more than a hundred Chinese journalists. Scenes flashed across a big screen: the ball, the boy, the giant, the dunk. Little Fatty looked adorable. Li Weichong, China Unicom's marketing director, gave a speech. "In America, people talk about the Ming dynasty," he said. "What does this mean? Now that Michael Jordan has retired, the NBA needs another great player. Our Yao Ming could be the one." The press conference ended with an instrumental version of the theme from *Titanic*.

South Korea and China played for the title on National Day—the fifty-fourth anniversary of the founding of Communist China. Ha Seung-Jin, the eighteen-year-old, came out inspired: after false-starting the jump ball, he immediately collected four points, two rebounds, one block, and a huge two-handed dunk. He also committed four fouls in less than four minutes. For the rest of the game, Ha sat on the bench, shoulders slumped.

The Chinese starting point guard fouled out in the third quarter, and then the backcourt began to collapse. The Korean guards tightened the press, forcing turnovers and hitting threes: Bang, Yang, Moon. Bang three, Bang three, layup—and with five minutes left China's lead had dwindled to one point.

On every possession, Yao came to half-court, using his height and hands to break the press. At one point, he dove for a loose ball—all seven feet six inches sprawled across the hardwood. With the lead back at five, and less than two minutes left, Yao grabbed an offensive rebound and dunked it. Thirty points, fifteen rebounds, six assists, five blocks. After the buzzer, when the two teams met at half-court, Yao Ming shook Ha Seung-Jin's hand, touched his shoulder, and said, "See you in the NBA."

\* \* \*

THE NEXT MORNING, YAO CAUGHT THE FIRST FLIGHT OUT OF Harbin. He sat in the front row of first class, wearing headphones. First the Indian team filed past, in dark wool blazers, and then the Filipinos, in tricolor sweat suits. The Iranians were the last team to board, Darrehsari's head scraping the ceiling. Each player nodded and smiled as he walked past Yao. During the flight, it seemed that all the Chinese passengers came forward to have their tickets autographed. In three days, Yao would leave for America. Later that month, he would accept an apology from Coca-Cola and settle the lawsuit out of court.

I sat in the row behind Yao, beside a chubby man in his forties named Zhang Guojun. He had flown to Harbin to watch the game, and his scalped ticket had cost nearly two hundred dollars. Zhang was proud of his money—he showed me his cell phone, which used China Unicom services and had a built-in digital camera. Zhang told me that he constructed roads in Inner Mongolia. He sketched a map on the headrest: "This is Russia. This is Outer Mongolia. This is Inner Mongolia. And this"—he pointed to nowhere—"is where I'm from."

We talked about basketball. "Yao is important in our hearts," Zhang said solemnly. "He went to America, and he returned." Halfway through the flight, the man held up his cell phone, aimed carefully, and photographed the back of Yao Ming's head.

# THE HOME TEAM

THE NIGHT BEFORE THE OPENING CEREMONIES OF THE 2008 Olympics, Wei Ziqi joined two of his neighbors on the local barricade. It consisted of a rope stretched taut across the road, and the villagers carried wooden paddles that read "Stop!" in both Chinese and English. Two of the neighbors wore blue-and-white polo shirts with the "Beijing 2008" logo across the breast. Sancha, their village, is a ninety-minute drive from the capital, and marks a point where the Great Wall winds through the mountains of northern China. At the barricade there was also a piece of paper with a message in English: "Please help us to protect the Great Wall. This section of the Great Wall is not open to the public."

According to the Beijing Organizing Committee for the Olympic Games, or BOCOG, there were more than 1.7 million citizen volunteers in the region. The most visible ones were stationed at Olympic events and at places like the airport and downtown intersections, which were usually staffed by high school and college students who spoke some English. These urban volunteers had been outfitted by Adidas, an official Olympic sponsor; the company provided gray trousers, new running shoes, and bright-blue shirts made of a high-tech material called ClimaLite. But the ClimaLite and the corporate sponsorship disappeared in the countryside. That was one way to gauge distance—north of the capital, the urban development thinned out, and along the way the volunteers' gear became more ragged. The ClimaLite was replaced by cheap cotton; the running shoes were no longer standard issue; the Adidas logo was nowhere to be seen. Many farmers wore only a red armband, because they were saving the new shirt for something more important than the Olympics.

And yet these rural volunteers were diligent. Sancha's population was less than two hundred, but the village had enlisted thirty residents to staff the barrier around the clock. Earlier that afternoon, when Wei Ziqi drove me through the countryside to the village, we were stopped at two other checkpoints. We also passed a crumbling Ming dynasty tower manned by a lonely sentinel wearing a green armband that read "Great Wall Groundskeeper." In Bohai township, six miles from the village, I registered with the police. For the Olympic period, the authorities had banned foreigners from spending the night in this part of the countryside, but they made an exception because I had rented a house in Sancha since 2001. "Just don't hike up to the Great Wall," the cop warned me. He said that the big tourist sites were open, but everything else was off limits. On his desk was a stack of police manuals titled *The Terrorist Prevention Handbook*. While we chatted, I opened one to a random chapter: "What to Do if There's a Terrorist Attack in a Karaoke Parlor."

For China, 2008 had already been the most traumatic year since 1989, when the Tiananmen Square massacre occurred. In March there had been riots in Tibet, followed by a brutal crackdown by the authorities. Overseas, human-rights demonstrators disrupted the Olympic torch relay, leading to an angry nationalist backlash in China. In May, a powerful earthquake in Sichuan Province killed more than sixty thousand people. Recently, there had been a fatal attack on Chinese military police in Xinjiang, a region in the far west where much of the native Muslim population resents China's rule. All these events had contributed to the stress of the Olympic year, but I didn't understand the concern about the Great Wall. "They're worried about foreigners, people who might want Tibet independence," Wei Ziqi told me. "They don't want them to go up to the Great Wall with a sign or something."

It was fear of a photo op—that somebody would unfurl a political banner and take a picture atop China's most distinctive structure. The government also worried that a foreigner might hike in a remote area and get injured, creating bad press. For this, the authorities had mobilized more than five thousand people in the region, but labor is plentiful in rural China. And these volunteers were getting paid. That was another difference from the city, where patriotic students were willing to donate their time to the Motherland's Olympic effort. Farmers were too practical for that; in addition to the free shirt, each rural volunteer received five hundred yuan a month, about seventy-three dollars. In Sancha, where the average resident earned about a thousand dollars per year, it was good money.

For Wei Ziqi, though, the Olympics didn't represent a windfall. He and his wife ran one of the few businesses in the village, a small restaurant and guesthouse, and now they missed the Chinese city dwellers who usually drove out on weekends. Since July 20, the government had restricted the use of private cars in an attempt to improve the capital's notorious air pollution. The system was regulated through license-plate numbers: cars with plates ending in even digits could be used only on even-numbered days, and the odds were limited to odd days. This effectively ended overnight trips—if anybody drove to the village and stayed past midnight, he was stuck there for another twenty-four hours.

I never heard Wei Ziqi complain about the Olympics; nor did he show any animosity toward the hordes of "Free Tibet" protestors who were supposedly threatening the Great Wall. For a middle- or upper-class Beijing resident, the reaction would have been more emotional—such people were proud of hosting the Games, and many Chinese had been upset about the disruptions of the torch relay. But rural folks knew the limits of what they could control, and there was a detachment from outside affairs, even those which affected the village. No adults in Sancha planned to attend any Olympic festivities. When I asked Wei Ziqi if he and his family would accompany me to some events, he said, "I don't want to go."

"Why not?"

"We're not supposed to go into the city," he said. "They don't want a lot of people there right now."

I assured him that in fact spectators with tickets were welcome at the Games.

"It's not necessary," he said. "We can watch it on television."

The night before the opening ceremonies, I joined him at the barricade with the other villagers. Wei Ziqi's shift was 9 P.M. to 6 A.M. Only two cars passed through, both of them dropping off locals. Afterward, the vehicles turned around to hustle back to the city, because they had odd-numbered plates. It was like *Cinderella*—nobody wanted to be on the road when the clock struck midnight.

One barricade volunteer was a middle-aged man named Gao Yongfu. "President Bush just arrived," he announced, fiddling with a small radio. "He's in Beijing now. Putin is coming, too." He continued, "An American company has the rights for Olympics television. They have the rights for the whole world! Even if China wants to broadcast the Games, they have to go through that American company."

The third volunteer, a woman named Xue Jinlian, didn't think that sounded right. "They can't control what China broadcasts," she said.

"Yes, they can."

"I don't think so. Not in China's own country!" Xue was silent for a while. "Chinese people are naturally smart," she finally said. "Their problem is they don't have enough money. You look at America, and a lot of the top scientists are Chinese. There's a lot of smart people here, but if there's not enough money, they leave."

Village conversations had a way of veering off suddenly, like a hawk that catches some invisible air current, but inevitably they returned to settle on certain topics: food, weather, money. Gao brought us back to the weather—the air was heavy, but he said the government wouldn't let it rain the following night. "They can make it rain somewhere else instead," he said. "I don't know how it works, but it uses high technology."

They were still discussing the weather a little before midnight, when I walked back to my house and went to bed. Later, I heard that the first car of August 8 had come through the barricade at around 2 A.M. The license plate ended with the number two—a Beijing motorist determined to make the most of his twenty-four hours. That same day, the government fired over a thousand silver iodide–laced rockets into the sky, ensuring perfect dryness for the opening ceremonies. At 5 A.M., jet lag woke me up and I wandered back down the road. Behind the barricade, Wei Ziqi dozed in the passenger seat of his car, and the morning light shone on the Jundu Mountains, and nothing about the peaceful scene suggested that this day was different from any other.

EARLIER THAT WEEK, I HAD ARRIVED ON UNITED FLIGHT NO. 889, San Francisco to Beijing. The airline was a sponsor of the U.S. Olympic team, and the gate in San Francisco had a pre-race feel: the solidarity of the starting line, that moment before the pack breaks into its inevitable divisions. The U.S. women's softball team was there, as were the synchronized swimmers. The American track cyclists had gathered near the windows that overlooked the tarmac. Two athletes from Belize wore matching green-and-black tracksuits. There was a National Olympic Committee member from Venezuela, an elderly man who wore a brown bow tie and carried a cane. The television people were easy to pick out—tall blond women with BlackBerrys and burnished skin. Jim Gray, a prominent sports announcer who was covering the Olympics for NBC, strode up and down the terminal, avoiding eye contact with anybody who recognized him.

Once boarding began, the solidarity dissolved. The TV folks vanished into first class and business class, along with the Venezuelan committee member. The pair from Belize sat quietly in economy. Most of the American athletes were in economy plus. The softball team sat along the left wing, and the cyclists and the synchronized swimmers took the right; if it wasn't perfect ballast, it was close. "Everybody here at United would like to welcome all the athletes," the pilot announced on the intercom, after takeoff. Later, he spoke again. "I want to pass on a message from the women's softball team," he said. "They want to say, 'Good luck to the men in tights!'"

There was no rejoinder from the cyclists. They wore white compression tights beneath their T-shirts and warm-ups, and periodically each athlete stood up and took a lap through the aisles, shaking out his legs. That was the United 889 velodrome: walk to the bathroom, turn at the exit row, duck past the softball mockery, and head back toward Belize. On one of these circuits, I interrupted Michael Friedman, who was scheduled to compete in men's track cycling in two weeks. He was a friendly twenty-five-year-old with reddish hair and a barrel chest. "It's for the blood clots," he explained, when I asked about the compression tights and the pacing. "We're not supposed to sit down for too long."

The flight took more than thirteen hours, and at Beijing airport's Terminal 3 we were greeted by smiling volunteers in ClimaLite. There were also representatives from United Airlines, who distributed information sheets to all American athletes. Among other matters, they warned the Olympians about the volunteers:

> We ask that you stay with the group and our staff as we've experienced instances where BOCOG volunteers (in blue uniform) with the intention to assist has directed part of our groups off on their own.

Other bullet-point instructions had a slightly ominous tone:

- Please note that Immigration officers normally will not provide an explanation when they take your passport and OIAC card away.

The athletes had fallen silent, and they gathered in a tight cluster, like cattle on the open range before a storm. Four cyclists wore black face masks, which covered the nose and mouth, and had the sharp lines of armored visors. Michael Friedman said that the masks had been issued by the team, in case the pollution in Beijing was bad. "They told us we should do this," he said. He looked a little sheepish; none of the athletes from other sports wore masks. But then they hadn't used the compression tights, either. "I figure, no reason to take any risks," Friedman told me with a shrug.

The cyclists wore their masks through baggage claim and customs. At the exit, television cameras were waiting, and the images created a brief uproar after they appeared. Within a day, the athletes had issued an apology through the United States Olympic Committee. It read, in part, "Our decision was not intended to insult BOCOG or countless others who have put forth a tremendous amount of effort to improve the air quality in Beijing." The day that I sat on the village barricade, the apology made the front-page headlines of the *China Daily*:

TORCH TIME IN TOWN AS FEVER RISES

PUTIN PRAISES PREPARATIONS

US CYCLISTS SORRY FOR WEARING MASKS

THE MEN'S ROAD-CYCLING FINAL WAS THE FIRST DAY AFTER THE opening ceremonies, and it was one of the few events that didn't require a ticket. The race began downtown, winding through the city before heading north to the Great Wall. Down the street from the Lama Temple, white metal barricades had been erected along the sidewalks. The ClimaLite crew was there, stationed at intervals of thirty feet, and there were also local volunteers in "Capital Public Order Worker" T-shirts. Plainclothes cops worked the crowd. In China, undercover officials have a distinct look: well-built men in their thirties or forties, dressed in button-down shirts, dark trousers, and cheap leather loafers. They almost always have crew cuts. They move in packs, and they linger and loiter; they have a tendency to stare. Their purpose is to intimidate as much as infiltrate: there's no need for subtlety in a one-party state. At the cycling race, the plainclothes cops had been issued little Chinese flags as a halfhearted attempt at cover, but they didn't wave them like everybody else. They held the flags beside their hips, like weapons at the ready.

On the sidewalk, two men played the game of Chinese chess known as *xiangqi*. They sat on stools around a wooden board, and they paid no attention to the growing mob of people. If they noticed the plainclothes men, they made no sign—long ago Beijing residents had learned to take surveillance in stride. And this was the chess players' turf: in the shade of a scholar tree, in front of the Badaling Leather Shoe Shop. Zhang Yonglin, one of the players, owned the shop. His opponent was a retired auto mechanic named Zhang Youzhi. The players were unrelated and locals referred to them as Little Zhang and Old Zhang. Forty minutes before start of the cycling race, a volunteer told them to leave.

"Wait until we finish this game," Old Zhang said.

He carried a fan inscribed with gold calligraphy, and he gestured with it, a brief swipe that indicated that the game wouldn't last long. The volunteer was lower caste—no ClimaLite—and she shrugged and left the men alone. A few minutes later, an official BOCOG volunteer walked over. "You need to move," he said. "There's going to be a bicycle race here."

"We know that," Old Zhang said. "We're just going to finish the game."

This time the flip of the fan was more dismissive. The young volunteer seemed reluctant to challenge this elderly man, and so the game continued. By now, seven people had gathered to watch, and one of them told me that Old Zhang was the best player in the neighborhood. In China, chess is a sport: the Chinese Xiangqi Association is administered by the All-China Sports Federation, just like the Chinese Cycling Association and the Chinese Basketball Association. The federation also handles bridge, go, darts, and the Chinese Tug-of-War Association. If this seems a muddled view of athletics, then it helps to think of the All-China Sports Federation as being concerned with competitive pastimes, broadly speaking. Beneath this umbrella organization, some associations exist with the primary goal of competing with foreigners at the Olympics. This is why the Chinese excel at obscure sports, and why so many of their 2008 gold medals would be gained in events that average citizens almost never encounter: archery (one gold), sailing (one), shooting (five), weight lifting (eight). They won a gold in canoeing, a form of water transport as Chinese as the tomahawk. It's a triumph of bureaucracy, and it shouldn't surprise anybody. If a nation can organize 1.7 million volunteers, from Tiananmen Square to the Great Wall, all of them outfitted according to subtle distinctions of class and status, then surely it should be possible to find and train one woman capable of winning the RS:X windsurfing gold. (Her name is Yin Jian.)

But like the concept of bureaucracy, chess had a presence in China long before medal counts. And Chinese chess truly feels like a sport, and so does chess-watching. It even has set positions. There's always at least one observer who gives advice before a move is made. Another onlooker waits until the move is finished before he offers his comments. This is the pairs event for spectators—the coach and the critic—and you would expect it to drive players to violence. But all aggression is directed at the board. Near the Lama Temple, Old Zhang and Little Zhang slammed the wooden pieces as hard as they could with every move.

*Thwack!*

"I'm giving your horse something to eat!"

*Thwack!*

"I need a gate! I need a gate!"

"Right, right! That's the right move!"

*Thwack!*

"I'm giving it to you cheap!"

With twenty-four minutes left until the cycling, and after the players had been asked to leave on three separate occasions, Little Zhang finally conceded the match. He did this Beijing style: he dumped the pieces on the ground and howled, "Old Zhang plays black!" Then they immediately began another game. By now, they had an audience of fifteen people, including four security volunteers in uniform. Periodically, a plainclothes man wandered over, flag at the hip, to watch for a few minutes.

As Old Zhang played, he used his fan like a master. He folded it when thinking, and after a move he always unfurled it with a flourish. Near the end of the game, when it became clear that he had left himself vulnerable, the fan began to move jerkily, as if in irritation; but still the old man said nothing. At last, he conceded with a smile. There were fewer than ten minutes left when the men finally put away the board.

Now the crowd pressed toward the barricades, and for a long time the road was empty. "They're coming!" somebody finally said.

"Cars are coming!"

"They're all Volkswagens," somebody else observed. The advance vehicles were black VW sedans with tinted windows. Then came a police motorcycle and police car, followed by a truck with a big platform that swiveled like a gun.

"That's the television camera!"

Two lead cyclists whizzed past, a Chilean and a Bolivian. Half a minute later the entire pack went by so fast that the crowd could hardly react. Nobody had any idea who was in front; the uniforms weren't printed in Chinese; the cyclists' faces were a blur. For an instant, there was stunned silence, and then everybody saw the long line of support cars and cheered.

"Why do they have the bikes on top?"

"That's for fixing them."

"Each one has a flag—look!"

"Those aren't Volkswagens, though."

"They're Skodas, I think."

"Skodas, definitely."

"There's an ambulance!"

They gave the last-place ambulance a good sendoff. For a few minutes, the street was empty, and then it was as if another race began. The leader was a battered bicycle cart carrying scraps of wood. A normal bike followed, then a Honda taxi. A truck full of bottled water. A parade of odd-numbered plates: 1, 7, 5, 9. The crowd dispersed; volunteers dismantled barricades; Old Zhang shuffled off to lunch. "It was OK," he said, referring to the bicycle race. Earlier, he had showed me his fan's calligraphy, which consisted of a poem titled "Do Not Get Angry."

"It reminds me to stay calm when playing chess," he said. The opening verses read:

> *Life is just like a play, and we are here only because of destiny*
> *It's not easy to be together until we're old, so why not cherish it?*

WHEN I TOLD WEI JIA I HAD AN EXTRA TICKET TO THE FENCING competition, he asked me which kind. Wei Jia was Wei Ziqi's eleven-year-old son, a native of rural Sancha. "There's *peijian, zhongjian,* and *huajian,*" he said matter-of-factly—sabre, épée, and foil. "The swords have different sizes and shapes."

Like all elementary school children in the greater Beijing region, Wei Jia had been issued a textbook called *The Primary School Olympic Reader.* It began in Olympia ("The grass is green and the flowers fragrant"), descended to cartoons of naked Greeks wrestling, and continued to Baron de Coubertin. One section featured the Finnish runner Paavo Nurmi; another was devoted to John Akhwari, a Tanzanian marathoner who, in 1968, showed great sportsmanship in finishing last. The chapter about Liu Xiang, China's great hurdler, made you wish the book had been issued with some wood to knock:

> Liu Xiang is healthy, and while training and racing he rarely gets injured, which is hard for an athlete.

It was partly Wei Jia's interest that persuaded his father to accept my offer to attend some events. The first one was rowing, and Wei Ziqi called the night before with a question about raincoats. "Do they give them out free?" he asked.

"I don't think so," I said. "Why would they do that?"

"I was watching on TV," he said. "Everybody in the stands has the same color raincoat."

Somehow, despite hours of watching the coverage, I had missed that detail. I said that the coats were probably being sold at venues, but Wei Ziqi was shrewder. "You're not allowed to use umbrellas, right?"

This was true, because of security concerns.

"Well, if they don't let people use umbrellas," he said, "then maybe they give out raincoats."

I didn't quite understand the logic, but the following day, after we passed through security at the Shunyi Olympic Rowing-Canoeing Park, about twenty miles outside of Beijing, the first thing we saw was a woman handing out cheap plastic ponchos. It was like that at every event—the organizers knew their crowd. The Chinese love freebies, and there were always volunteers distributing something: plastic flags, cheap cardboard binoculars, fans with the McDonald's logo. They gave out pamphlets that described the rules of various sports and told the audience how to behave. (At volleyball: "Applauding is welcome at appropriate times during the match. Booing and jeering is not allowed.") Concessions were unbelievably cheap. They sold instant noodles, which Chinese like to eat dry, for thirty cents. A cold can of beer cost less than seventy-five cents. It was possible to show up at fencing at 10 A.M., spend four and a half bucks, and get a six-pack of Budweiser. But no Chinese person would ever do that. Few Chinese have spent much time at spectator sports, and there's no tradition of drinking at the ballpark. Most people buying beer seemed to be foreigners.

The Chinese were focused and they were intense. There was none of the looseness of the street; these people had paid for their tickets, and they knew the opportunity wouldn't come again. They were often silent at the beginning of an event, almost on edge as they tried to figure out the action; it got loud later, especially after Chinese athletes appeared. At the preliminaries of men's sabre, during the first hour, a fistfight broke out three rows ahead of Wei Jia and me. It was like a play within a play: in the background, Renzo Agresta of Brazil was slashing at Luigi Tarantino of Italy, and then two Chinese men stood up and started whaling on each other. They appeared to be middle class; one was accompanied by his child. In China, public disagreements are common, and typically they consist of a lot of shouting that goes nowhere. But at fencing there was no prelude and no encore, just ten seconds of flailing. By the time the ClimàLite showed up, nobody was talking, and the volunteer couldn't figure out what had happened. The men simply sat down and shut up, terrified of getting kicked out. The most I learned from neighbors was that the dispute had started over sight lines.

The woman next to me was named Wang Meng, and she was a graduate student in agriculture. She had received her ticket from a friend, who bought it online more than a year earlier. At face value, it cost $4.40; when I asked Wang if she would have resold it for $300, she shook her head. "This is my only chance to see something at the Olympics," she said. She spent the first half hour talking in hushed tones with her neighbors, trying to figure out what it means when a fencer's helmet lights up.

In the past, I had disliked attending athletic events in China, because the nationalism can be so narrow-minded. Few people care about the sport itself; victory is all that matters, and there's little joy in the experience. But at the Olympics I sensed something different. Cao Chunmei, Wei Jia's mother, talked mostly about how the events made her feel. "It's peaceful," she said at synchronized diving, whereas wrestling made her nervous. She said the Bird's Nest—Beijing National Stadium—was *luan*, "chaotic." "That's the way it should be," she said. "A real nest is like that." Her favorite was the National Aquatic Center, the Water Cube. When I commented that the patterned exterior resembled bubbles, she disagreed: "Those are too big to be bubbles." She liked the place because it gave her a clean feeling.

The brand-new rowing park in Shunyi also initially made

her nervous, because she couldn't swim and disliked boats. (When I asked Wei Ziqi if he could swim, he said, "A little.") There was a brief rain burst and the family sat happily in their free ponchos. Rural people travel light—none of the Weis had brought anything with them, not even Wei Jia, who planned to stay with me in the city to see more of the Games. After the event, we said good-bye to his parents and got in a cab. I asked the driver to recommend a restaurant in Shunyi.

"The Golden Million is good," he said. Shunyi is about twenty miles outside Beijing, and it's the kind of small city that's common in the suburbs of sprawling Chinese municipalities. Many Shunyi residents are former peasants on their way to becoming something else, and local officials were proud to host rowing, canoeing, and kayaking. They had hung banners all around town that said "Culture is in Shunyi / The Olympics are in Shunyi." In the city center, the Golden Million Restaurant's mirrored entrance was decorated with 493 bottles of Old Matisse Scotch. At the center of the restaurant was an enormous tank filled with a dozen sharks, two soft-shelled turtles, and one woman dressed as a mermaid. In addition to a long mono-fin, she wore a bikini top, a face mask, and a nose clip. A sign said: "THE CAPITAL'S TOP MERMAID SHOW!" The tank was circular and the woman swam laps with the sharks and turtles. One advantage of traveling with Wei Jia was that he often came up with the questions that I was too stunned to ask.

"Why is that lady in the water?" he said, looking concerned, when the waitress came to our table.

"It's a type of performance," she said.

"Why don't the sharks bite her?"

"Because they're full," she said, smiling reassuringly. "If you feed them, they don't bite people."

Later Wei Ziqi told me that back at the rowing complex, after I got in the taxi with Wei Jia, the next cabbie had refused to take him and his wife. There was a long line of cars, but the drivers had been instructed to take foreigners only: Chinese spectators had to wait for the public bus. Wei Ziqi laughed when he told the story; he didn't take it personally.

In February of 2001, when Beijing was bidding to host the Games, I had accompanied the IOC inspection commission on its last tour of the capital. For more than three hours, our motorcade traveled through the city, visiting potential stadium sites; everywhere we went, the traffic lights turned green, as if by magic. (The day before, the Chinese had demonstrated how all signals could be changed by remote in the city's traffic-control center.) Along the inspection route, the facades of hundreds of buildings had been freshly painted in bright colors. According to government statistics, workers had brushed up twenty-six million square meters, which meant that they had painted an area nearly half as big as Manhattan.

At that time, even dissidents spoke out in favor of the bid, hoping that the Olympics would bring political change. That possibility was reportedly a factor in the IOC's decision; many members believed that the 1988 Games, in Seoul, had helped to reform South Korea. "What the hell is the party going to do?" an American who had served as an IOC adviser told me, back in 2001. "It's going to be really hard to have a Stalinist party and open the Olympic Games." Liu Jingming, Beijing's vice-mayor, told me that the organizers had considered but finally rejected the slogan "Great Wall, Great Olympics." Seven years later, it was clear that the Communist Party could indeed open the Games, and the decision regarding the slogan seemed wise, given that there were now some five thousand farmers guarding the structure against foreigners.

Ever since Deng Xiaoping, China had become steadily more receptive to the outside world, but there was still an element of fear and insecurity. The Olympics undoubtedly helped to promote openness, but the Games didn't initiate a political transformation, just as they didn't change the basic outlook of most people. Long ago, the Chinese had learned how to take big events in stride; this was how they survived Mao, and it was how they weathered disasters like the Sichuan earthquake. And such fortitude was evident throughout the Games, although you had to know where to look. You could see it in the calmness of the divers, the toughness of the weight lifters, the discipline of the gymnasts. Most Chinese athletes come from poor rural areas where children are recruited into sports schools. Sancha was already too prosperous for that—no village kids in recent memory had attended a sports school, and Wei Ziqi told me that he wouldn't allow his son to take such a route. But other places had fewer options, and parents were satisfied to have their children enter the well-structured bureaucracy of the sports system.

The farmers had also left their mark on the facilities, at least indirectly. When I reread the promotional materials that BOCOG had given me at the time of the bid, the sketches of proposed arenas looked modest in comparison with what was eventually built. This was the opposite of the typical pattern— usually, an Olympic host city promises the moon and then scales back. In 2001, the Chinese government said that they would build six subway lines spanning 140 kilometers; they ended up with eight lines of 200 kilometers. The sketches of proposed arenas looked blocky and bland and utilitarian. There was no Bird's Nest, no Water Cube, nothing of distinction. But since then the economy had boomed, driven by the continued migration of people from the countryside, who also provided the labor for the elaborate construction—this time it was more than just paint. In a sense, China had reached the perfect stage for hosting the Games: workers were still cheap, political accountability remained at a minimum, and the rising middle and upper classes could attend and take pride in the competition. They were the ones who responded most deeply to the Olympics—crowd energy came largely from the young and the affluent. In the stands, it was easy to forget that most Chinese are still country people.

AT EVENTS, I LIKED WANDERING THE DISTANT CORNERS OF THE arenas. Often the worst seats had been given away: at men's sabre, in the most remote section, a hundred and fifty Beijing Forestry Bureau workers sat with thunder sticks in hand, looking slightly dazed. At the preliminaries for Greco-Roman wrestling, there was a group of schoolchildren from Changping, a city outside Beijing. Their teacher stood up to make an announcement: the father of a competitor was sitting right behind them!

The man was in the very last row. His name was Chang Aimei, and he was fifty-two but he looked at least a decade older. He had dark skin and sun-wrinkled eyes, and he carried a fieldworker's white sweat towel. In his lap he clutched all of the day's freebies: a Chinese flag, a flag with the Olympic mascot, an English guide to the Games. He also had a Chinese pamphlet with instructions on how to behave at Greco-Roman wrestling: "Spectators are encouraged to greatly applaud wrestlers who have demonstrated superior skills, or who have scored high points."

Chang Aimei's son was named Chang Yongxiang, and moments earlier he had defeated the reigning world champion, a Bulgarian. Chang Aimei was calling his wife; like many people from the countryside, he shouted every time he handled a cell phone. "Eldest Son just competed!" he yelled. "He won! What? I said he won!"

For the next hour, the phone rang repeatedly: relatives, friends, reporters from back home. The Changs came from Hanba, a village of fewer than three thousand people in Hebei Province. They farmed wheat and corn on three-quarters of an acre of land, and their annual income was less than six hundred dollars. In the 1980s, one of Chang Aimei's nephews had been selected for a wrestling program, eventually becoming Chinese national champion. After that, Chang Aimei believed there were opportunities for his boy, who was naturally big. At the age of thirteen, the son left home to enter the county sports school, and then moved to the provincial level and the national team. He now wrestled in the seventy-four-kilogram weight division. I asked Chang Aimei why he sat in the last row.

"The coaches don't want him to know I'm here," he said. "They don't want anything to disrupt him, so they told me to sit in the back."

It was only the second time he had seen his son wrestle. Attending the Olympics had been a surprise; three days earlier, local officials had told him that they had some tickets and he could have one. The other tickets went to the village party secretary and the director of the county sports bureau, both of whom had moved down to better seats. Chang Aimei's daughter waited outside the arena; she hadn't been able to get a ticket.

The cell phone rang again. "An American reporter is interviewing me!" he shouted. "Right now! American!"

Two years ago, his son had competed at a meet in Colorado Springs. "He said you Americans are really nice," Chang Aimei told me. "He said it's very clean there, too."

It was early in the day and the athletes were working their way through the rounds. At the Olympics, Chinese men had never done better than a bronze in wrestling; nobody had ever made the finals. In Chang Yongxiang's second match of the morning, he defeated a Peruvian to qualify for the semis. When it came time for his next competition, I made my way to the far corner of the arena. His father was still there, sitting alone.

Chang Yongxiang was matched against a Belarusan named Aleh Mikhalovich. All morning the crowd had grown louder, and now they chanted: "China, go! China, go! China, go!" Almost immediately the Belarusan threw Chang out of the ring, scoring four points, and he won the first period. But then Chang seemed to gather himself. He was stocky, with thick thighs and a square jaw. He had bristly black hair, and after every clinch he shook his head like a bull. In the second period, he evened the match, and now the spectators were on their feet. The school group from Changping screamed and banged their thunder sticks.

Behind them, Chang Aimei remained seated. His legs were crossed, as if he were relaxing after a day's labor, and his belongings were neatly stacked on his lap: towel, flags, pamphlets. He had not moved a muscle since the match began. His eyes were fixed on the distant mat, and he said nothing. But I could hear him breathing—steady, steady, steady. In the third period the Belarusan took the initial point. Deeper now, deeper now. The match continued with Chang Yongxiang in the lower position; he escaped and scored a point. Inhale—almost a gasp. Another point, and then it was over, and the referee was raising Chang Yongxiang's arm.

Eventually, Chang lost to a Georgian, taking the silver medal. But on the day of the semifinal he left the ring triumphant, already the most successful Chinese Greco-Roman wrestler in history. The crowd roared—*China, go! China, go!* At the top of the arena, safely out of sight, Chang Aimei still looked relaxed. He was silent until he took out the cell phone. "*Wei!*" he shouted. "He just won again!"

# CAR TOWN

EVERY DAY, THE AMERICANS VISITED THE AUTO MANUFACTURER in Wuhu. There were twenty of them: engineers, company executives, market specialists, technical consultants. One lawyer. The auto manufacturer was called Chery, and it was a new company that had been booming over the past two years. On most mornings, the American engineers test-drove prototypes on a small track outside Chery's final-assembly plant. The test track had signs in Chinese; the engineers spoke Detroit.

"I'd give it a six."

凹凸路

"Pay attention to the clutch engagement."

制动检测

"Do you remember the Versailles?"

"Of course."

"What an embarrassing car."

请保特车距40米

"The only nine I've ever given was the idle on a Lincoln Marquis."

毛石路

"Are you going to do a J-turn? 'Cause if you do, warn me."

The leader of the American team was Malcolm Bricklin, who had founded a company called Visionary Vehicles to partner with Chery. When I met him in Wuhu, he shook my hand and announced that they were going to become the first company to import Chinese-made cars to the United States. He said, "We're making history, and I'm going to film it." His son was the videographer. Jonathon Bricklin was in his twenties, and he followed his father everywhere he went, with the tape rolling.

Malcolm Bricklin was sixty-six years old, and he had spent most of his life chasing breakthroughs in the automobile industry. In the late 1960s, he introduced Subaru to American consumers and made a small fortune. In the early 1970s, he spent much of that fortune founding an automobile manufacturer in New Brunswick, Canada. He commissioned the design of a futuristic sports car with gull-wing doors, named it after himself, and promptly went bankrupt.

In the 1980s, Bricklin brought the Yugo across the Atlantic. He declared personal bankruptcy not long after that. Later he tried to make electric bicycles in California, but Americans would never love bikes the way that they love cars.

In 2002, he began searching for a way to get back into the auto business. He had the idea that he'd find another foreign manufacturer that was capable of exporting cars to the United States. He visited England, Serbia, Romania, Poland, and India. He stopped looking once he got to Wuhu.

This was Visionary Vehicles' second trip to the Chinese city, and they were staying at the Guoxin Hotel. Each morning, they ate breakfast in the executive lounge, and Malcolm Bricklin held forth about the past and the future. He was a tall man with white hair and striking gray-blue eyes, and his voice was deep and smooth. He never sat still. He claimed that in Wuhu he had stumbled upon the perfect car company. At breakfast he was usually joined by Tony Ciminera, the executive vice president of Visionary Vehicles, and Ronald E. Warnicke, who was the company's lawyer. Warnicke was an old friend of Bricklin's, and one of his specialties was bankruptcy law. In his garage, back in Arizona, he still had a Bricklin SV-1 with gull-wing doors.

"People talk about the Yugo like it was some kind of big failure," Bricklin said. "Tony's job back then was to find the least expensive car in the world. That was when Yugoslavia was a Communist country—but a Communist country that was friendly to the West. We found a fifteen-year-old car that had never had to meet any regulations. It was basically a Fiat 128."

Tony Ciminera said, "We had to bring our own cases of toilet paper, our own fax toner. We had to bring unleaded gas for the cars."

Bricklin said, "Henry Kissinger was our consultant. Tony made five hundred and twenty-eight changes in the car in fourteen months, and in fourteen months the car was in dealers' hands."

Tony said, "We built a factory outside of their factory, just to fix every car that came out. And in the States the car was selling so fast that dealers charged three thousand over the price, which was thirty-nine hundred."

Bricklin said, "GM moved up Saturn into a more expensive market niche, because of the Yugo. And the quality was starting to get better. We were selling a lot of cars, and we were getting a lot of acclaim. Three years later I sold out my share. And then the war started. Now everybody says that the Yugo was a failure."

The Guoxin Hotel, like everything else in this part of Wuhu, was brand-new. The bookcase of the executive lounge had been filled with books the way you would stock a freshly dug pond with fish. There were twelve copies of *Harvard Marketing Management* in Chinese and ten copies of *M.B.A. Harvard Business School Management Encyclopedia*.

Bricklin said, "We're not bringing in a cheap Chinese car. We're bringing in cars of value and selling them for less. A twenty-thousand-dollar car will go for fourteen; a thirty-thousand-dollar car for twenty. We're talking about 30 percent of the market." He shifted restlessly, eyes flashing. "I see the similarities with Japan," he said. "In 1968, it was the turning point—where people's perceptions of Japanese products shifted from cheap to good quality." He continued, "What the Japanese took twenty years to do, the Chinese can do in five."

A COUPLE OF DAYS BEFORE BRICKLIN AND THE OTHERS HAD arrived in Wuhu, I had driven there from Beijing. The journey was eight hundred miles; I rented a Chinese-made Volkswagen Jetta and took my time. The road passed through Confucius' hometown, and it also went by the Stone Warriors of Nanpi, the Iron Lion of Shijia, and the Alfalfa Land of Jinniu. I drove past the Dongguang Iron Buddha and the Wuqiao Acrobatics World. All of them were advertised by big signs that promoted the local specialty. The village of Jinxiang had erected a huge billboard that said, in English: "The Best Garlic Is from Jinxiang in All of China."

The highway was excellent—four lanes, groomed medians, well-marked exits. Some sections were so new that they still appeared on my map as broken lines. The Chinese expressway network had doubled in length in the past four years, and recently the ministry of communications had held a press conference to announce plans to add another thirty thousand miles. When asked about the purpose of the new roads, Zhang Chunxian, the minister, mentioned Condoleezza Rice's visit to the People's Republic during the previous year. Apparently, Rice had told a Chinese official about the fond memories she had of summer vacations spent traveling in the family car. "She said those trips helped her love the United States," Zhang explained. "By building expressways, we can boost the auto industry, but that's only a small part of it."

Along the way to Wuhu, I drove past miles of shiny new billboards; they were as blank as unplugged televisions, waiting for advertisers to figure out what kind of consumers might someday pass this way. In recent years, increasing numbers of urban Chinese had purchased automobiles, but it was still rare for them to take long trips, because tolls were high and drivers inexperienced. Most of the other vehicles on the road were transport trucks. That was the first stage of a Chinese highway: first they moved objects, and later the people would arrive.

The truckers worked in teams of two or three, so they could drive in shifts around the clock. They followed regular routes, and they owned their rigs; any delay cost money and they ate fast at roadside restaurants. Every night, I stopped for a meal and whatever conversation I could catch. Occasionally, I met a trucker with a poetic streak—one man described his trade as "the thermometer of the economy"—but usually we had just enough time to establish the basics before they rushed back out to the parking lot. One pair told me that they had a truck full of bamboo whisk brooms; they had just dropped off a shipment of nonferrous metal. Another pair had unloaded color televisions and picked up processed wheat. These were the alchemists of the new economy, at the center of every mysterious exchange that occurred along the Chinese highway system. One pair of truckers had just dropped off computerized mah-jongg sets and picked up elementary school textbooks. Another team had exchanged radiators from Hangzhou for chemical materials from Shijiazhuang. Shoes from Wenzhou; electric generators from Changchun. Coal from Datong; train parts from Wenzhou. Nobody drove an empty rig.

In the city of Qufu, I pulled over at the Confucius family cemetery. The local government had turned it into a tourist attraction, with signs posted at the highway exit, but my car was the only one in the lot. The cemetery sprawled through a big forest; for more than twenty centuries, local men with the Confucius family name—Kong—had been buried here, along with their wives. The numbers were staggering: a hundred thousand people had been laid to rest among the cypress trees.

I wandered aimlessly, looking at tombstones. I saw one that dated to the late Ming dynasty and the sixty-second Kong generation; I took a few steps to the next memorial and skipped three hundred years: 2001, seventy-fourth generation. I was on my way to another tablet when I heard wailing. I followed the sound, picking my way around the tombs.

A group of women cried and kowtowed beside a mound of fresh earth. They had arrived on a Taishan 200 tractor, which had three wheels and a two-stroke engine. The tomb offerings were simple: oranges, apples, a boiled chicken. The men stood nearby, watching, and one of them offered me a cigarette. He told me that they were burying a woman who had been married to a member of the seventy-second generation of Kongs. The wailing continued for ten more minutes and then it stopped as abruptly as if a meter had run out. Two women came over to chat; they asked me what American funerals were like, and what my salary was, and if it was true that people in the States could have as many children as they wanted. I told them that I was of the fifth generation of Hesslers in America. A man crank-started the tractor and they puttered off into the mist. They left the chicken but packed up the oranges.

Nearby, the stone memorial above Confucius' tomb was still cracked from the vandalism of the Cultural Revolution. A tour guide said that Red Guards had excavated the grave and found it empty. He smiled when he told the story; I couldn't tell if his point was that the vandals had been thwarted or that Confucius had never been buried here in the first place. When I left, the parking lot was still vacant, and so was the highway. The only difficult moment of the trip occurred south of Tianjin, where traffic suddenly slowed and swerved, because drivers were distracted by hundreds of pamphlets that flapped above the road like dying birds. I pulled over and caught one. It was in English: a fourteen-page mortgage application for Woolwich, a financial-services company in Dartford, Kent. Apparently, a truck full of imported recycled materials had come unlatched on the road. There were thousands of the forms, fluttering in the air and skidding beneath tires, and they were as blank as all the billboards.

THERE WERE PIECES OF CARS SCATTERED EVERYWHERE AROUND the world, and Malcolm Bricklin tried to find ways to connect them. He searched out bankruptcies and remote factories, because this was often where the opportunities could be found. One morning at breakfast, I asked him to explain in more detail how he had come to Wuhu.

"Three years ago, I got a call from a contact in Yugoslavia," he began. "He asked if I would please come and see them, because they wanted to give me a factory. This is a factory that NATO put five missiles in. We went there and spent a year. But the problem was that they have all of these workers, and what do you do with those employees while you're fixing the plant? Then the prime minister got assassinated, and we said, We're too old for this.

"We went to Romania, where there was a Daewoo plant. But it was the same problem—what do you do with the employees? We went to Poland, great place, a fabulous Daewoo factory. They were selling engines to the Ukraine. Don't ask why. Same problem, though—all these employees. But they introduced us to MG Rover, which was interested in working with the Poland plant. There were too many uncertainties, though. We went to Tata, in India. Really nice people."

"Very nice," said Tony Ciminera.

"We go to see the factory, and the factory is OK," Bricklin said. "It's not state of the art. They have one model coming out, and they are oversold. We're trying to figure out which pieces of the puzzle make sense. A Russian we had met, he had come to us for engines to sell to Central America. He said, Why aren't you going to China? I said, I don't feel like another trip. He said,

You ought to go see them, they are really smart, really aggressive. Then he said, They're right in Shanghai, why don't you go see them? We decided to do it. And then right before we left, he said, Oh, you'll have to take a train to the town. Thank God he lied. So we arrive and take a train to Wuhu, and it's completely packed; we're sitting like this for five hours."

For the first time all morning, Bricklin is motionless—he pulls his arms close to his body and freezes, as if pinned between other passengers. Then he jumps back to life: "They took us to see the plant, and we were blown away. I pull out a letter of intent and hand it to them. And then we spent seven hours dealing with that. They said they wanted to build a relationship; I said I'm not flying back and forth to build a relationship. That night we met the president for dinner. Everything he was telling us was exactly what you'd want. So we signed a letter of intent. It took forty-eight hours to do the letter."

THERE WAS A TIME WHEN THE CHINESE ELITE DISDAINED BUSINESS. Traditional Confucian values meant that any educated person scorned the merchant class, and emperors rebuffed the first Western attempts at trade. But the British were determined to buy tea and sell opium, and they were willing to fight to do so. In 1842, after the first Opium War, the Chinese were forced to sign the Treaty of Nanking, which opened five cities to British trade. It established a pattern: if the Chinese weren't willing to open their markets willingly, you could always find a pretext and turn to violence. In 1858, after more fighting, the Chinese agreed to designate ten more ports for foreigners. In 1876, after a British consul was murdered by tribesmen in western China (he was scouting possible trade routes to Burma), the Qing government agreed to open four more treaty ports; and one of them was Wuhu.

The city sits on the east bank of the Yangtze, in the land-locked province of Anhui. In the late 1870s, the British built a pillared consulate on a hill above town, and beside the river they set up a customs house, where opium was processed. French Jesuits constructed a church; the Spaniards ran a Catholic school. American Protestant missionaries built a hospital. And then the twentieth century brought a new series of events—the fall of the Qing dynasty, the Japanese invasion, the Communist revolution—and the foreigners disappeared from Wuhu. During the decades of the Communist planned economy, the central government invested little in the region.

After 1978, when China entered the period known as Reform and Opening, one of Deng Xiaoping's key strategies involved export-processing zones—designated areas that encouraged foreign investment through special tax rates. Many of these early zones happened to be located in the former treaty ports, and like the old days, they appeared in waves. In 1992 and 1993, after the earliest zones like Shenzhen had thrived for more than a decade, the central government approved thirty-two more. One of them was Wuhu.

The city was a latecomer to the new economy, and it was relatively remote. There was no distinctive local product of real value. During the 1980s, Wuhu had briefly become known for producing a brand of sunflower seeds known as "Idiot Seeds" (it rhymes in Chinese), and that was even worse than getting stuck with Alfalfa Land or the Best Garlic in All of China. Wuhu's leaders wanted a genuine core industry, and they sensed that their obscurity could actually be an advantage in a heavily regulated industry like car manufacturing. Since the 1980s, foreign automakers had been allowed to form joint ventures with state-owned Chinese partners, with foreign ownership limited to 50 percent. The government's goal was to build an industry quickly, learning from the foreigners while maintaining control. (In China, nobody forgot the humiliation of the treaty ports.) Companies like Volkswagen and General Motors joined with Chinese partners, manufacturing foreign-brand cars while farming out parts production to cheap China-based suppliers. For a time, the strategy was extremely profitable for everybody, in part because the strict government controls limited competition.

In Wuhu, though, the officials quietly skirted the regulations. They recruited Yin Tongyao, an Anhui native and a trained engineer, who had been a rising star in one of the joint ventures with Volkswagen. Yin had helped move some of the tooling and equipment of a failed VW plant from Westmoreland, Pennsylvania, to Changchun, Jilin. In Westmoreland, the factory had produced Golfs and Jettas. In Changchun, they used the same platform—the car's basic frame and major components—to build a Chinese version of the Jetta, which eventually became the best-selling car in the country.

Yin left Volkswagen to become the vice president of Wuhu's new company. With funding from the local government, he bought manufacturing equipment from an outdated Ford engine factory in England and shipped it to Wuhu. Next, he went to Spain, where he acquired manufacturing blueprints for a model called the Toledo, which was being produced by a struggling Volkswagen subsidiary called SEAT. The Spanish car used the same platform as the Jetta.

In secret, Yin built an automobile assembly line in Wuhu. National regulations forbade new auto manufacturers from entering the market, so the officials in Wuhu named their enterprise an "automotive components" company. The first engine came off the line in May of 1999. Seven months later they turned out a car. It used Jetta parts, acquired from suppliers who were supposedly exclusive to Volkswagen. VW was furious, and so was the central government.

But this was a common strategy in the Reform era: push the boundaries first, then ask forgiveness. For more than a year, Wuhu's officials negotiated with the central government, finally receiving permission to sell their cars nationwide in 2001. (Reportedly, Volkswagen accepted a financial settlement and decided not to sue.) In Chinese, they named their company Qirui, which has connotations of good fortune. It's pronounced a little like "cheery," but the English name was spelled without the second *e*, to indicate that Chery would always be one step removed from the complacency that comes from happiness. By 2004, Chery was selling almost ninety thousand vehicles a year.

When I met with Chu Changjun, the Party deputy secretary of the Wuhu Economic and Technological Development Area, I asked why Chery hadn't followed the proper channels for establishing a car company.

"It's like having a child," he explained. "First you have a child, then you register him. We did it the same way: First we made the cars, and then we registered to make cars."

Chu was joined by He Xuedong, the deputy director, who added, "If you apply in the traditional manner, you'll end up waiting for years. During that time, the opportunity may very well pass you by."

We were in the development area's new Investment Services Center, whose marbled lobby was big enough for two badminton nets to be stretched side by side. Some cadres had a pretty good game going when I visited during lunch break. Chu and He gave me a stack of promotional materials in English, which included the mottoes "An Oasis for Investors" and "Investors Are Our God." One sentence read, "In Wuhu, we have high-quality human resource with low cost." The next sentence discussed electricity, water supply, and sewers.

The development area was north of town, on the banks of the Yangtze. With Chery as an anchor tenant, Wuhu had already attracted more than a hundred manufacturers, and new plants were going up all the time. Many of the factories had been moved from the southern coast, where costs and wages had increased during two decades of strong economic growth.

One afternoon, I visited Baoshun Road, in Wuhu's second

factory district, which was lined with plants in various stages of construction. Empire Hill planned to start production the following month (air-conditioner electrical parts); Shuncheng Electronics was about to add another hundred workers (air-conditioner wiring). Century made four thousand plastic air-conditioner covers a day. At the Wuhu Shijie Hardware Company, when I asked about their products, the gatekeeper opened a desk drawer, grabbed a handful of shiny new nails, and tossed them down like dice. A pretty young woman appeared and handed me a business card that read "Merry Yeh—International Trade Dept. Manager." Merry served me tea in an upstairs office, where a plaque commemorated the company's cooperation with National Nail Corporation ("More Than Just Nails") in the United States. Wuhu Shijie produced sixty tons of nails a day. Three hundred and forty workers. Eighteen thousand square meters. Any other questions?

"Where did you get your English name?"

"I named it after Merry Christmas!"

Downtown, the old British consulate had been converted into the local Communist Party headquarters, and the Spanish missionary school had been turned into a technical college. For years, the former customs house, which once processed opium, had served as a kindergarten, but now it was abandoned. You could still see the faded Cultural Revolution slogans that had been painted vertically onto the hated symbol of British gunboat diplomacy. But small-time traders had erected wooden structures around the base, cutting off the slogans:

*Chairman Mao's Writings Are—*
*In the Midst of Struggle One Learns—*

ONE AFTERNOON, AN AMERICAN NAMED JOHN DINKEL DROVE a Chery T-11 prototype out into the Wuhu development zone. Visionary Vehicles had hired Dinkel as a technical consultant, and his specialty was road testing. "You find out how good a car is when you do bad things," he told me, as we pulled out of the factory. Three young Chinese test engineers sat in the back. None of them wore seat belts.

The T-11 was a sport-utility vehicle, which eventually would be released on the Chinese market as the Tiggo. It looked suspiciously similar to the Toyota RAV4. Dinkel also planned to test a new crossover wagon known as the B-14, which was scheduled to hit Chinese dealers later that year. Chery was upgrading its domestic product line, in part to prepare for more demanding international markets. The cars destined for the United States were still in the early design-and-development stage, and one of the most formidable hurdles would be meeting American safety and emissions standards. Even though the T 11 and the B 14 wouldn't be exported, Dinkel planned to evaluate the quality while showing the Chery engineers how Americans road-tested vehicles. He had asked me to come along and help with translation.

On a street in the development zone, he ran through a series of checks: accelerating, braking, turning. "It's picking up a wheel," he said, in the middle of a tight turn. "The wheel is spinning. You need a limited slip differential for that." I did my best with the translation, Detroit to Chinese. We passed a tractor cart full of bricks, a new air-conditioner factory, and a boy pissing in the grass. Dinkel sped up and swerved; a bus honked. In the back, the three engineers clutched at the ceiling. Finally one of them asked me to pass along a request: "Do you think we could go to a place without any other cars?"

That was easy: in China, there was always another, newer development zone just down the road. We drove north, past bulldozers, earthmovers, and the skeletons of future housing complexes, to Baoshun Road. Dinkel said: "Tell them the gearbox is very notchy from second to third, and from fourth to fifth."

Dinkel was alert, soft-spoken, and small-framed. As a graduate student at the University of Michigan at Dearborn in the late 1960s, he was the only guy in the emissions laboratory who could fit into the driver's seat of a Mazda Cosmo. When I asked why he had originally studied engineering, Dinkel said, "I didn't have a very bright guidance counselor." In those days, people believed that anybody with good math scores should become an engineer. After briefly working at Chrysler, he switched to journalism. He was at *Road & Track* for twenty years, including two as editor-in-chief. "I've tested cars for thirty years," he said. "I've driven practically every car that's ever been on the road." He told me that Wuhu's empty streets reminded him of the old days in California, when they could still test cars in the bean fields of Orange County.

At the west end of the factory strip, between the nail factory and the air-conditioner-cover plant, stood an empty roundabout. To John Dinkel, it looked a lot like a skid pad. He accelerated to forty miles an hour, past a tepee of stacked bamboo that would be used to scaffold the next construction project. He held the turn, tires squealing, and the roundabout flashed by: Nails, bamboo, air-conditioner covers. Nails, bamboo, air-conditioner covers. The three Chinese engineers were thrown together against the right side of the car. They still weren't wearing seat belts.

The one in the middle was named Qi Haibo. He could have fit into the driver's seat of a Mazda Cosmo, along with a sack of groceries. He was twenty-two years old, and he had grown up on a farm in Inner Mongolia. His grandfather had moved there from Shaanxi Province ("probably because of famine or war"). Qi Haibo's father had a fifth-grade education and his mother had attended one year of elementary school. They grew wheat, corn, and sunflowers.

As a boy, he had always been the top student in his primary and middle schools. After high school, despite having no special interest in engineering, he attended Wuhan Polytechnic

University. "I wanted to go to a good university, and I heard that computers and electronics were the best subjects for careers nowadays," he said. "So I chose those specialties when I took the examination."

A year ago, during his last term in college, he attended a job fair and met some Chery recruiters. "They offered me a job, and people at the school said it was a new company, a company that was developing fast. So the next day I signed a contract. I figured that a young person could learn a lot there." By Chery standards, he wasn't particularly young—the average age of a company employee was twenty-four. Qi worked six days a week for a salary of less than two hundred dollars per month. He lived in a factory dorm room with three other engineers. He would have preferred his own space, but the dorm conditions were a lot better than anything he'd known in Inner Mongolia, and he hoped for a long-term future at Chery. "I also like the fact that it's not a joint venture," he said. "It's China's own auto company."

After the test drive, I asked Qi Haibo what he had learned from John Dinkel. Qi said that the T-11 had a slight problem with driveshaft length, which meant that the outside wheel slipped a little on tight turns. The rear end of the B-14 had a tendency to float at high speeds. In particular Qi admired Dinkel's skills behind the wheel. The Chinese engineer, whose job involved quality control and test-driving, had received his license only one month earlier.

CHERY'S EARLY STRATEGY—GLEANING USEFUL INFORMATION from the wreckage of troubled automakers around the world— culminated in a tiny vehicle called the QQ. In the 1990s, Daewoo Motors, the South Korean company, tried to expand rapidly around the world, investing in factories in Vietnam, India, Poland, Romania, Ukraine, and Uzbekistan. Soon, they realized that they had overextended themselves—Uzbekistan, it turned out, wasn't a great place for a car factory—and Daewoo declared bankruptcy. The big American firms took their time picking through the wreckage, watching the prices drop. In 2002, after studying the situation for more than a year, General Motors finally acquired a controlling interest in Daewoo Motors. GM took the platform of a Daewoo minicar known as the Matiz, repackaged it under the name Chevrolet Spark, and prepared to start production in China.

In June of 2003, half a year before the Spark went on the market, Chery unveiled the QQ. It looked almost exactly like GM's car but retailed for a quarter less: about six thousand dollars. Chery also introduced a sedan that appeared suspiciously similar to the Daewoo Magnus. Chery named that car the Son of the Orient.

Chinese consumers never took to the Son of the Orient, but the QQ was an instant success. At less than twelve feet long, and with a 0.8-liter engine, it was even smaller than the Mini Cooper. The car was perfect for China's fledgling urban middle class, which hadn't been able to afford the expensive joint-venture products in the past. In 2003, nationwide passenger-car sales jumped by 80 percent, allowing Chery and other small manufacturers to establish a strong position in the low-end market. In 2004, Chery sold five times as many QQs as GM sold Sparks.

In December of 2004, GM Daewoo filed a lawsuit in Shanghai, alleging that Chery had developed the QQ "through copying and unauthorized use of GM Daewoo's trade secrets." Simple copyright violations were rampant in China, but this case was more complicated: Chery had essentially bootlegged a car before the original even made it to market. The implication was that top-secret designs had leaked out of South Korea, probably during the year when GM had negotiated to acquire the bankrupt company.

When I visited the GM China offices in Shanghai, Timothy P. Stratford, the company's general counsel, handed me two photographs. In the first picture, two cars were parked side by side: the green one was the QQ, and the black one was the Matiz, the South Korean original. In the second picture, the doors had been switched: green on black, black on green.

"You would never find two competitors' cars where the doors could be swapped," Stratford explained. "It means that not only do they copy the door but everything else that is necessary to form the opening for the door. A door opening is kind of like a fingerprint for a car."

Chery executives hadn't made many public statements about the case, apart from saying that they had received Chinese patents for the QQ (a point that would be moot if designs had been obtained illegally). When I spoke with Lin Zhang, the general manager for Chery's International Division, he emphasized that the QQ was already on the market when he joined the company. But he denied that there had been any wrongdoing, and said that it was natural for a young company like Chery to legally develop something similar to a car that had been successful elsewhere.

"That's how you get started, at a primitive level, and then you move to another level," he said. "It's like when you start drawing. You don't begin by drawing a beautiful picture of your own—you imitate another picture. It's in the nature of any industry. It's how Sony and Hyundai and Toyota got started. They all started with something. And it's something they abandon quickly."

Zhang had joined Chery a year earlier. Born in Shanghai, he had a PhD in mechanical engineering from the University of Michigan at Ann Arbor. He had worked for nine years for DaimlerChrysler in Detroit. He had two children, aged eight and ten, and they had been happy in American schools. But every year, when Zhang visited Shanghai, there was some new part of town that he didn't recognize, which made him think about how much he was missing. After a friend—another American-trained Chinese engineer—accepted a job at Chery, Zhang followed suit.

"If I stayed, I could see what I'd be doing in five or ten years," he told me. "The life will be relatively easy but it's missing excitement. I thought that risk and reward usually come together."

Risk was part of Chery's culture, and even the QQ may have been a calculated gamble. Several independent analysts told me they doubted that the Chinese legal system was sophisticated enough to handle such a case. In addition, Chery was government-owned, and the Chinese had always dreamed of having a true national brand. GM Daewoo hoped to have its day in court, but there was no guarantee that the foreign company would be treated fairly. And since Chery didn't intend to export the QQ to the United States, there wasn't much recourse in the American legal system.

Chery was still relatively small—eight thousand employees— but it had found a way to produce a hundred thousand vehicles a year without significant investment in design and engineering. Management appeared to be shifting strategies; an enormous R&D center had just been completed, and they had recently hired almost thirty foreign-trained engineers, like Zhang. The company was placing new emphasis on quality control, and every expert who visited their factories seemed impressed with the scale and sophistication. Ronald E. Harbour, an American consultant who specializes in auto manufacturing, told me that Chery's aluminum-casting plant was so big that they were only using 10 percent of the space. "In China, they tend to build a lot of capacity ahead of demand," he said. "They seem to have an endless supply of money. I don't know where it comes from. Most Western companies would not be able to throw in that kind of money on the if-come."

As a state-owned company, Chery didn't have to answer to stockholders, and it was unclear how much had been invested. One day, I visited the gigantic final assembly plant that produced the QQ and the Son of the Orient. A sign at the entrance said, in Chinese:

## WE NEED NOT ONLY HARD WORK,
## WE MUST ALSO BE DILIGENT,
## AND MORE IMPORTANTLY WE MUST HAVE A
## SENSE OF A NATIONAL MISSION

Nearby, a digital sign noted that they were producing 253 QQs a day. Workers in blue uniforms moved the new cars along the assembly line. A manager named Hu Bin told me that over the past couple of years they had steadily picked up the production pace, to meet growing demand. In the beginning, assembly-line stations in Hu's section had spent an average of three minutes performing their tasks; now they required only two minutes and five seconds. Hu Bin said soon Chery would start paying workers by the car instead of by the hour. He asked me if American manufacturers sometimes used that tactic to boost production; I said no, and mentioned the word "union." Hu hoped that his section would drop eighteen more seconds by the end of the year.

ANOTHER MORNING AT BREAKFAST IN THE GUOXIN HOTEL, Malcolm Bricklin said that Wuhu needed a new port on the Yangtze.

"They have to dig deeper, pave big streets, put in lighting for security and for inspecting damage," he said. "You have to be able to put five thousand cars on a ship in five hours."

In the United States, Bricklin was searching for dealers willing to pay up to $4 million each for the rights to sell Chinese cars. Visionary Vehicles also needed a new American brand name. "It's got to be something that fits but doesn't come from the name 'tiger' or 'dragon,'" Bricklin said. "I happen to like the name Chery." Not long after this conversation, GM's lawyers sent Bricklin a warning to the effect that Chery, besides being one letter away from Cheery, was also one letter away from Chevy and another possible lawsuit. And not long after that, GM and Chery finally settled the Spark-QQ dispute out of court, with neither side making public comment about the terms.

Such tensions were common between American and Chinese companies, who were still figuring out how to coexist in the same world. When I spoke to Michael Dunne, the president of the consulting firm Automotive Resources Asia, in Shanghai, he wondered if the appeal of Chery's low cost would be significant enough to distinguish a new brand in America. There was a history of foreign automakers—Hyundai, for instance—stumbling upon their first exposure to American consumers, usually because of poor quality. Toyota had succeeded because of its meticulousness—something that Detroit lacked for a long time. Dunne believed that the Chinese shared Detroit's flaw. "The Chinese are a little like Americans," he said. "They want the touchdown. They want the home run. *Women hen congming.*"

That was a phrase the Chinese sometimes said: "We are very clever." Dune continued, "But I don't see the patience and perseverance. It's more like: 'We can leap-frog.'"

I understood what he meant: the longer I lived in China, the more certain aspects reminded me of America. There was the same boundless optimism and energy; both the Americans and the Chinese built wide roads across instant cities. They often had the quality of an upstart, and they believed that they could defeat time—in China, that characteristic sometimes seemed even more American than the Americans. Whenever the Chinese engineers stood beside the Visionary Vehicles crew, the men from the younger country suddenly looked old: jet-lagged, gray-haired, paunchy. They had already been around the block—Bricklin had been a millionaire and a bankrupt; he had been acclaimed as a visionary and condemned as a fraud. He talked so smoothly in part to keep the past at bay.

At the end of 2006, Bricklin's relationship with Chery collapsed. Two years later, Visionary Vehicles filed a lawsuit against Chery, for an estimated $40 million in damages. Bricklin began chasing the next thing, plug-in hybrid vehicles. And Chery continued to grow, becoming China's biggest car exporter. Over the years, they agreed to partnerships with Fiat and Jaguar Land Rover. But the vast majority of their exports were directed at the developing world—by 2012, they still hadn't created a vehicle for the American market.

When I was in Wuhu, I met with Yin Tongyao, the former VW employee who had risen to become president of Chery. He was known for avoiding the press, but late one evening he agreed to meet me and a few other reporters in a conference room at the Guoxin Hotel. He wore a suit and tie, as if he'd just left the office. He was in his early forties and looked younger. But when I asked about his education, he spoke as if describing ancient history.

"At the time I left for college," he said, "I had never even ridden in a passenger car before." He explained that in those days the top graduates were assigned to jobs in truck factories, because China didn't have a market for passenger cars. He had been a bad student, he said with a self-deprecating laugh, and thus he was sent off to the northeast. I asked about the company's strengths.

"All we have is our aggression," he said. "We have no brand name, no recognition, nothing. We are simply aggressive."

I COULDN'T SLEEP THE LAST NIGHT IN WUHU. THE LONG DRIVE to Beijing lay ahead, and light and noise filtered into my room. At 4:30 A.M., I finally got up and looked out the window. Across the street, the Conch factory hummed with overtime labor; its workers were producing PVC window frames. The Visionary Vehicles team had left the day before. I was the last American left in the Guoxin Hotel.

I walked outside and started the Jetta. A mist hung low over the development zone; my headlights cut across empty streets. The Chinese New Year was approaching, and many factories were working overtime in preparation for the holiday. The blocky buildings were illuminated from within, like paper lanterns.

It was dawn when I crossed the new Yangtze bridge and got on the expressway. A few miles outside Wuhu, I passed an exit sign for a place called Wuwei. It was an ancient Daoist phrase: Do Nothing. In this region there was another city called Wuxi: No Tin. Once, in Sichuan Province, I took a bus through Asbestos. That was a tough name for a small city, but it was better than having nothing at all.

# CHINESE BARBIZON

In the countryside southwest of Lishui, where the Da River crosses a sixth-century stone weir, the local government announced that it was founding a Chinese version of the Barbizon. The original French Barbizon School developed during the first half of the nineteenth century, in response to the Romantic movement, among painters working at the edge of the Fontainebleau Forest. Back then, the French artists celebrated rural scenes and peasant subjects. This wasn't exactly the mood in Lishui—like most cities in eastern China's Zhejiang Province, the place was focused on urban growth; there was a new factory district and the export economy was booming. But the local Communist Party cadres wanted the city to become even more outward-looking, and they liked the foreign cachet of the Barbizon. They also figured it would be good business: art doesn't require much raw material, and it's popular overseas. They referred to their project as Lishui's *Babisong*, and they gave it the official name of the Ancient Weir Art Village. One party slogan described it as "A Village of Art, a Capital of Romance, a Place for Idleness."

In order to attract artists, the government improved some old riverside buildings and offered free rent for the first year, with additional subsidies to follow. Painters arrived immediately; soon the village had nearly a dozen private galleries. Most people came from China's far south, where there was already a flourishing industry of art for the foreign market. Buyers wanted cheap oil paintings, many of which were destined for tourist shops, restaurants, and hotels in distant countries. For some reason, the majority of artists who settled Lishui's Barbizon specialized in cityscapes of Venice. The manager of Hongye, the largest of the new galleries, told me that it had a staff of thirty painters, and that its main customer was a European-based importer with an insatiable appetite for Venetian scenes. Every month he wanted a thousand Chinese paintings of the Italian city.

Another small gallery, Bomia, had been opened by a woman named Chen Meizi and her boyfriend, Hu Jianhui. The first time I met Chen, she had just finished a scene of Venice, and now she was painting a Dutch street scene from what looked like the eighteenth century. A Russian customer had sent a postcard and asked her to copy it. The painting was twenty inches by twenty-four, and Chen told me she would sell it for about twenty-five dollars. Like everybody in the Ancient Weir Art Village, she referred to Venice as Shui Cheng, "Water City," and Dutch scenes were Helan Jie, "Holland Street." She said that over the past half year she had already painted this particular Holland Street as many as thirty times. "All the pictures have that big tower in it," she said.

I told her that it was a church—the steeple rose in the distance, at the end of a road bordered by brick houses with red-tile roofs.

"I thought it might be a church, but I wasn't sure," she said. "I knew it was important because whenever I make a mistake they send it back."

Through trial and error, she had learned to recognize many of the landmark buildings of Europe. She had no idea of the names of St. Mark's Basilica and the Doge's Palace, but she knew these places mattered, because even the tiniest mistake resulted in rejection. She worked faster on less iconic scenes, because customers didn't notice slight errors. On the average she could finish a painting in under two days.

Chen was in her early twenties, and she had grown up on a farm near Lishui; as a teenager, she learned to paint at an art school. She still had a peasant's directness—she spoke in a raspy voice and laughed at many of my questions. I asked her which of her pictures she liked the most, and she said, "I don't like any of them." She didn't have a favorite painter; there wasn't any particular artistic period that had influenced her. "That kind of art has no connection at all with what we do," she said. The Barbizon concept didn't impress her much. The government had commissioned some European-style paintings of local scenery, but Chen had no use for any of it. Like many young Chinese from the countryside, she had already had her fill of bucolic surroundings. She stayed in the Ancient Weir Art Village strictly because of the free rent, and she missed the busy city of Guangzhou, where she had previously lived. In the meantime, she looked the part of an urban convert. She had long curly hair; she dressed in striking colors; she seemed to wear high heels whenever she was awake. On workdays, she tottered on stilettos in front of her easel, painting gondolas and churches.

Hu Jianhui, Chen's boyfriend, was a soft-spoken man with glasses and a faint crooked mustache that crossed his lip like a calligrapher's mistake. Once a month, he rolled up all their finished paintings and took a train down to Guangzhou, where there was a big art market. That was how they encountered customers; none of the buyers ever came to the Ancient Weir Art Village. For the most part, foreigners wanted Holland Streets and the Water Citys, but occasionally they sent photographs of other scenes to be converted into art. Hu kept a sample book in which a customer could pick out a picture, give an ID number, and order a full-size oil painting on canvas. HF-3127 was the

Eiffel Tower. HF-3087 was a clipper ship on stormy seas. HF-3199 was a circle of Native Americans smoking a peace pipe. Chen and Hu could rarely identify the foreign scenes that they painted, but they had acquired some ideas about national art tastes from their commissions.

"Americans prefer brighter pictures," Hu told me. "They like scenes to be lighter. Russians like bright colors, too. Koreans like them to be more subdued, and Germans like things that are grayer. The French are like that, too."

Chen flipped to HF-3075: a snow-covered house with glowing lights. "Chinese people like this kind of picture," she said. "Ugly! And they like this one." HF-3068: palm trees on a beach. "It's stupid, something a child would like. Chinese people have no taste. French people have the best taste, followed by Russians, and then the other Europeans. Americans are after that. We'll do a painting and the European customer won't buy it, and then we'll show it to a Chinese person, and he'll say, 'Great!' "

LISHUI IS A THIRD-TIER FACTORY TOWN, AND IN A PLACE LIKE that the outside world is both everywhere and nowhere at all. In the new development zone, assembly lines produce goods for export, but there isn't much direct foreign investment. There aren't any Nike factories, or Intel plants, or signs that say DuPont—important brands base themselves in bigger cities. Lishui has a central population of around two hundred fifty thousand, small by Chinese standards, and local companies make pieces of things: zippers, copper wiring, electric-outlet covers. The products are so obscure that you can't tell much from the signs that hang outside factory gates: Jinchao Industry Co., Ltd.; Huadu Leather Base Cloth Co., Ltd. At the Lishui Sanxing Power Machinery Co., Ltd., the owners have posted their sign in English, but they did so from right to left, the way Chinese traditionally do with characters:

## DTL, .OC YRENIHCAM REWOP GNIXNAS IUHSIL

It's rare to see a foreign face in Lishui. Over a period of three years, I visited the city repeatedly, talking to people in the export industry, but I never met a foreign buyer. Products are sent elsewhere for final assembly, some passing through two or three levels of middlemen before they go abroad; there isn't any reason for a European or American businessman to visit. But despite the absence of foreigners the city has been shaped almost entirely by globalization, and traces of the outside world can be seen everywhere. At a factory called Geley, workers make three-dollar plastic light switches that are marketed proudly as "The Jane Eyre Series." When Lishui's first gym opened, it was called the Scent of a Woman, after the Al Pacino movie. Once, I met a demolition-crew worker who had a homemade tattoo on his left arm that said "KENT." He told me he'd done it himself as a kid, after noticing that American movie gangsters have tattoos. I asked why he'd chosen that particular word, and he said, "It's from the cigarette brand in your country." Another time, I interviewed a young factory boss who wore a diamond earring in the shape of the letter $K$. His girlfriend had the O: whenever they were together, and the letters lined up, everything was all right.

The degree of detail impressed me. The outside world might be distant, but it wasn't necessarily blurred; people caught discrete glimpses of things from overseas. In many cases, these images seemed slightly askew—they were focused and refracted, like light bent around a corner. Probably it had something to do with all the specialization. Lishui residents learned to see the world in parts, and these parts had a strange clarity, even when they weren't fully understood. One factory technician who had never formally studied English showed me a list of terms he had memorized:

PADOMIDE BR. YELLOW 8GMX
SELLANYL YELLOW N-5GL
PADOCID VIOLET NWL
SELLAN BORDEAUX G-P
PADOCID TURQUOISE BLUE N-3GL
PADOMIDE RHODAMINE

In the labyrinth of the foreign language, he'd skipped all the usual entrances—the simple greetings, the basic vocabulary—to go straight to the single row of words that mattered to him. His specialty was dyeing; he mixed chemicals and made colors. His name was Long Chunming and his coworkers called him Xiao Long, or Little Long. He worked at Yashun, a factory that manufactured the tiny rings that connect to the adjustable straps of brassieres. It was a typically obscure Lishui product: a thin steel ring coated with nylon, weighing less than half a gram. The average bra includes four such rings, and the color has to match that of the other components. Whenever a bra assembly plant made an order with Yashun, they sent a sample strap, and Little Long studied the color. He would consult his notebook and figure out the right mixture of chemicals necessary to make Sellanyl Yellow or Padocid Turquoise Blue.

He had grown up on a farm in Guizhou, one of the poorest provinces in China. His parents raised tea, tobacco, and vegetables, and Little Long, like both his siblings, left home after dropping out of middle school. It's a common path in China, where an estimated hundred and fifty million rural migrants have gone to the cities in search of work. Little Long happened to find his first assembly-line position at a bra factory, and since then he had stayed in the industry, moving from job to job. Eventually he served as an assistant to a ring dyer, who taught him the trade. By the time I met Little Long, his entire professional life had been spent around brassieres, and this experience had given him one extremely specific subject about which he could speak in a worldly manner. "The Japanese like to have little flowers on their bras," he once told me, with the air of a connoisseur. "The Russians don't like that. They just want bras to be plain and brightly colored. And big!"

In the factory town, he had become relatively successful, earning a good wage of three hundred dollars a month. But he was determined to further improve himself, and he studied self-help books with foreign themes. In his mind, this endeavor was completely separate from his work. He had no pretensions about what he did; he never spoke of the export brassiere industry as giving him a special link to the outside world. As far as he was concerned, the skills he had gained were strictly and narrowly technical. "I'm not mature enough," he told me once, and he collected books that supposedly improved moral character. One was titled *The Harvard MBA Comprehensive Volume of How to Conduct Yourself in Society*. Another book was called *Be an Upright Person, Handle Situations Correctly, Become a Boss*. In the introduction, the author describes the divides of the worker's environment: "For a person to live on earth, he has to face two worlds: the boundless world of the outside, and the world that exists inside a person."

Little Long had full lips and high cheekbones, and he was slightly vain, especially with regard to his hair, which was shoulder-length. At local beauty parlors, he had it dyed a shade of red so exotic it was best described in professional terms: Sellan Bordeaux. But he was intensely serious about his books. They followed a formula that's common in the self-help literature of Chinese factory towns: short, simple chapters that feature a story about some famous foreigner and conclude with a moral. In *A Collection of the Classics*, the section on effective use of leisure time gave the example of Charles Darwin. (The book explains that Darwin's biology studies began as a hobby.) Another chapter told the story of how a waiter once became angry at John D. Rockefeller after the oil baron left a measly one-dollar tip. ("Because of such thinking, you're only a waiter," Rockefeller shot back, according to the Chinese book, which praised his thrift.)

Little Long particularly liked *A Collection of the Classics* because it introduced foreign religions. He was interested in Christianity, and when we talked about this subject he referred me to a chapter that featured a parable about Jesus. In this tale, a humble doorkeeper works at a church with a statue of the crucifixion. Every day, the doorkeeper prays to be allowed to serve as a substitute, to ease the pain for the Son of God. To the man's surprise, Jesus finally speaks out and accepts the offer, under one condition: if the doorkeeper ascends the cross, he can't say a word.

The agreement is made, and soon a wealthy merchant comes to pray. He accidentally drops a money purse; the doorkeeper almost says something but remembers his promise. The next supplicant is a poor man. He prays fervently, opens his eyes, and sees the purse: overjoyed, he thanks Jesus. Again, the doorkeeper keeps silent. Then comes a young traveler preparing to embark on a long sea journey. While he is praying, the merchant returns and accuses the traveler of taking his purse. An argument ensues; the merchant threatens to get the law; the traveler fears he'll miss the boat. At last the doorkeeper speaks out— with a few words, he resolves the dispute. The traveler heads off on his journey, and the merchant finds the poor man and retrieves his money.

But Jesus angrily calls the doorkeeper down from the cross for breaking the promise. When the man protests ("I just told the truth!"), Jesus criticizes him:

> What do you understand? That rich merchant isn't short of money, and he'll use that cash to hire prostitutes, whereas the poor man needs it. But the most wretched is the young traveler. If the merchant had delayed the traveler's departure, he would have saved his life, but right now his boat is sinking in the ocean.

When I flipped through Little Long's books, and looked at his chemical-color vocabulary lists, I felt a kind of vertigo. In Lishui that was a common sensation; I couldn't imagine how people created a coherent worldview out of such strange and scattered contacts with the outside. But I was coming from the other direction, and the gaps impressed me more than the glimpses. For Little Long, the pieces themselves seemed to be enough; they didn't necessarily have to all fit together in perfect fashion. He told me that after reading about Darwin's use of leisure time, he decided to stop complaining about being too busy with work, and now he felt calmer. John D. Rockefeller convinced Little Long that he should change cigarette brands. In the past, he smoked Profitable Crowd, a common cigarette among middle-class men, but after reading about the American oil baron and the waiter he switched to a cheaper brand called Hibiscus. Hibiscus were terrible smokes; they cost about a cent each and the label immediately identified the bearer as a cheapskate. But Little Long was determined to rise above such petty thinking, just like Rockefeller.

Jesus' lesson was the easiest of all: Don't try to change the world. It was essentially Daoist, reinforcing the classical Chinese phrase *Wu wei er wu bu wei.* ("By doing nothing everything will be done.") In Little Long's book, the parable of the crucifixion statue concludes with a moral:

> We often think about the best way to act, but reality and our desires are at odds, so we can't fulfill our intentions. We must believe that what we already have is best for us.

ONE MONTH, THE BOMIA GALLERY RECEIVED A COMMISSION TO create paintings from photographs of a small American town. A middleman in southern China sent the pictures, and he requested a twenty-four-inch-by-twenty-inch oil reproduction of each photo. He emphasized that the quality had to be first-rate, because the scenes were destined for the foreign market. Other than that, he gave no details. Middlemen tended to be secretive about orders, as a way of protecting their profit.

When I visited later that month, Chen Meizi and Hu Jianhui had finished most of the commission. Chen was about to start work on one of the final snapshots: a big white barn with two silos. I asked her what she thought it was.

"A development zone," she said.

I told her it was a farm. "So big just for a farm?" she said. "What are those for?"

I said the silos were used for grain.

"Those big things are for grain?" she said, laughing. "I thought they were for storing chemicals!"

Now she studied the scene with new eyes. "I can't believe how big it is!" she said. "Where's the rest of the village?"

I explained that American farmers usually live miles outside of town.

"Where are their neighbors?" she asked.

"They're probably far away, too."

"Aren't they lonely?"

"It doesn't bother them," I said. "That's how farming is in America."

I knew that if I hadn't been asking questions, Chen probably wouldn't have thought twice about the scene. As far as she was concerned, it was pointless to speculate about things she didn't need to know; she felt no need to develop a deeper connection with the outside. In that sense, she was different from Little Long. He was a searcher—in Lishui I often met such individuals who hoped to go beyond their niche industry and learn something else about the world. But it was even more common to encounter pragmatists like Chen Meizi. She had her skill, and she did her work; it made no difference what she painted.

From my outsider's perspective, her niche was so specific and detailed that it made me curious. I often studied her paintings, trying to figure out where they came from, and the American commission struck me as particularly odd. Apart from the farm, most portraits featured what appeared to be a main street in a small American town. There were pretty shop fronts and well-kept sidewalks; the place seemed prosperous. Of all the commissioned paintings, the most beautiful one featured a distinctive redbrick building. It had a peaked roof, tall old-fashioned windows, and a white railed porch. An American flag hung from a pole; there were flowers out front. A sign on the second story said "Miers Hospital 1904."

The building had an air of importance, but there weren't any other clues or details. On the wall of the Chinese gallery, the scene was completely flat: neither Chen nor I had any idea what she had just spent two days painting. I asked to see the original photograph, and I noticed that the sign should have read "Miners Hospital." Other finished paintings also had misspelled signs, because Chen and Hu didn't speak English. One shop called Overland had a sign that said "Fine Sheepskin Leather Since 1973"; the artists had turned it into "Fine Sheep-skim Leather Sine 1773." A "Bar" was now a "Dah." There was a "Hope Nuseum," a shop that sold "Amiques," and a "Residentlal Bboker." In a few cases I preferred the new versions—who wouldn't want to drink at a place called Dah? But I helped the artists make corrections, and afterward everything looked perfect. I told Chen that she'd done an excellent job on the Miners Hospital, but she waved off my praise.

Once, not long after we met, I asked her how she first became interested in oil painting. "Because I was a terrible student," she said. "I had bad grades, and I couldn't get into high school. It's easier to get accepted to an art school than to a technical school, so that's what I did."

"Did you like to draw when you were little?"

"No."

"But you had natural talent, right?"

"Absolutely none at all!" she said, laughing. "When I started, I couldn't even hold a brush!"

"Did you study well?"

"No. I was the worst in the class."

"But did you enjoy it?"

"No. I didn't like it one bit."

Her responses were typical of migrants from the countryside, where there's a strong tradition of humility as well as pragmatism. In the factory town, people usually described themselves as ignorant and inept, even when they seemed quite skilled. That was another reason that Chen took so little interest in the scenes she painted: it wasn't her place to speculate, and she scoffed at anything that might seem pretentious. As part of the Barbizon project, the cadres had distributed a promotional DVD about Lishui, emphasizing the town's supposed links to world art. But Chen refused to watch the video. ("I'm sure it's stupid!") Instead, she hung the DVD on a nail beside her easel, and she used the shiny side as a mirror while working. She held up the disk and compared her paintings to the originals; by seeing things backwards it was easier to spot mistakes. "They taught us how to do this in art school," she said.

Together with her boyfriend, Chen earned about a thousand dollars every month, which is excellent in a small city. To me, her story was amazing: I couldn't imagine coming from a poor Chinese farm, learning to paint, and finding success with scenes that were entirely foreign. But Chen took no particular pride in her accomplishment, and she talked about making art in the same way that Little Long described dyeing bra rings. These endeavors were so technical and specific that, at least for the workers involved, they essentially had no larger context. It was like taking their first view of another country through a microscope.

The Lishui experience seemed to contradict one of the supposed benefits of globalization: the notion that economic exchanges naturally lead to greater understanding. But Lishui also contradicted the critics who believe that globalized links are disorienting and damaging to the workers at the far end of the chain. The more time I spent in the city, the more I was impressed with how comfortable people were with their jobs. They didn't worry about who consumed their products, and very little of their self-worth seemed to be tied up in these trades. There were no illusions of control—in a place like Lishui, which combined remoteness with the immediacy of world-market demands, people accepted an element of irrationality. If a job disappeared or an opportunity dried up, workers didn't waste time wondering why, and they moved on. Their humility helped, because they never perceived themselves as being the center of the world. When Chen Meizi had chosen her specialty, she didn't expect to find a job that matched her abilities; she expected to find new abilities that matched the available jobs. The fact that her vocation was completely removed from her personality and her past was no more disorienting than the scenes she painted—if anything, it simplified things. She couldn't tell the difference between a foreign factory and a farm, but it didn't matter. The mirror's reflection allowed her to focus on details; she never lost herself in the larger scene.

WHENEVER I WENT TO LISHUI, I MOVED FROM ONE SELF-contained world to another, visiting the people I knew. I'd spend a couple of hours surrounded by bra rings, then paintings of Venice, then manhole covers, then cheap cotton glove liners. Once, walking through a vacant lot, I saw a pile of bright-red high heels that had been dumped in the weeds. They must have been factory rejects; no shoes, just dozens of unattached heels. In the empty lot the heels looked stubby and sad, like the detritus of some failed party—they made me think of hangovers and spilled ashtrays and conversations gone on too long.

The associations were different when you came from the outside. There were many products I had never spent a minute thinking about, like pleather—synthetic leather—that in Lishui suddenly acquired a disproportionate significance. More than twenty big factories made the stuff; it was shipped in bulk to other parts of China, where it was fashioned into car seats, purses, and countless other goods. In the city, pleather was so ubiquitous that it developed a distinct local lore. Workers believed that the product involved dangerous chemicals, and they thought it was bad for the liver. They said that a woman who planned to have children should not work on the assembly line.

These ideas were absolutely standard; even teenagers fresh from the farm seemed to pick them up the moment they arrived in the city. But it was impossible to tell where the rumors came from. There weren't any warnings posted on factories, and I never saw a Lishui newspaper article about pleather; assembly-line workers rarely read the papers anyway. They didn't know people who had become ill, and they couldn't tell me whether there had been any scientific studies of the risks. They referred to the supposedly harmful chemical as *du*, a general term that simply means "poison." Nevertheless, these beliefs ran so deep that they shaped that particular industry. Virtually no young women worked on pleather assembly lines, and companies had to offer relatively high wages in order to attract anybody. At those plants you saw many older men—the kind of people who can't get jobs at most Chinese factories.

The flow of information was a mystery to me. Few people had much formal education, and assembly-line workers rarely had time to use the Internet. They didn't follow the news; they had no interest in politics. They were the least patriotic people I ever met in China—they saw no connection between the affairs of state and their own lives. They accepted the fact that nobody else cared about them; in a small city like Lishui, there weren't any NGOs or prominent organizations that served workers. They depended strictly on themselves, and their range of contacts seemed narrow, but somehow it wasn't a closed world. Ideas arrived from the outside, and people acted decisively on what seemed to be the vaguest rumor or the most trivial story. That was key: information might be limited, but people were mobile and they had confidence that their choices mattered. It gave them a kind of agency, although from a foreigner's perspective it contributed to the strangeness of the place. I was accustomed to the opposite—a world where people preferred to be stable, and where they felt most comfortable if they had large amounts of data at their disposal, as well as the luxury of time to make a decision.

In Lishui, people moved incredibly fast with regard to new opportunities. This quality lay at the heart of the city's relationship with the outside world: Lishui was home to a great number of pragmatists, and there were quite a few searchers as well; but everybody was an opportunist in the purest sense. The market taught them that—factory workers changed jobs frequently, and entrepreneurs could shift their product line at the drop of a hat. There was one outlying community called Shifan where people seemed to find a different income source every month. It was a new town; everybody had been resettled there from Beishan, a village in the mountains where the government was building a new hydroelectric dam to help power the factories. In Shifan, there was no significant industry, but small-time jobs began to appear from the moment the place was founded. Generally, these tasks consisted of piecework commissioned by some factory in the city.

Once a month, I visited a family called the Wus, and virtually every time they introduced me to some new and obscure trade. For a while they joined their neighbors in sewing colored beads onto the uppers of children's shoes; then there was a period during which they attached decorative strips to hair bands. After that, they assembled tiny lightbulbs. For a six-week stretch, they made cotton gloves on a makeshift assembly line. (That job dried up, they told me, because unusually warm winter temperatures killed the market.)

On one visit to Shifan, I discovered that the Wus' son, Wu Zengrong, and his friends had purchased five secondhand computers, set up a broadband connection, and become professional players in a video game called World of Warcraft. It was one of the most popular online games in the world, with more than seven million subscribers. Players developed characters over time, accumulating skills, equipment, and treasure. Online markets had sprung up in which people could buy virtual treasure, and some Chinese had started doing this as a full-time job; it had recently spread to Lishui. The practice is known as "gold farming."

Wu Zengrong hadn't had any prior interest in video games. He hardly ever went online; his family had never had an Internet connection before. He had been trained as a cook, and he found jobs in small restaurants that served nearby factory towns. Occasionally, he did low-level assembly-line work. But his brother-in-law, a cook in the city of Ningbo, learned about World of Warcraft, and he realized that the game paid better than standing over a wok. He called his buddies, and three of them quit their jobs, pooled their money, and set up shop in Shifan. Others joined up; they played around the clock in twelve-hour shifts. All of them had time off on Wednesdays. For World of Warcraft, that was a special day: the European servers closed for regular maintenance from 5 A.M. until 11 A.M., Paris time. Whenever I visited Shifan on a Wednesday, Wu Zengrong and his friends were smoking cigarettes and hanging out, enjoying their weekend as established by World of Warcraft.

They became deadly serious when they played. They had to worry about getting caught, because Blizzard Entertainment, which owns World of Warcraft, had decided that gold farming threatened to ruin the game's integrity. Blizzard monitored the site, shutting down any account whose play pattern showed signs of commercial activity. Wu Zengrong originally played the American version, but after getting caught a few times he jumped over to the German one. On a good day he made the equivalent of about twenty-five dollars. If an account got shut down, he lost a nearly forty-dollar investment. He sold his points online to a middleman in Fujian Province who went by the Internet name Fei Fei.

One Saturday, I spent an afternoon watching Wu Zengrong play. He was a very skinny man with a nervous air; his long, thin fingers flashed across the keyboard. Periodically, his wife, Lili, entered the room to watch. She wore a gold-colored ring on her right hand that had been made from a euro coin. That had become a fashion in southern Zhejiang, where shops specialized in melting down the coins and turning them into jewelry. It was another ingenious local industry: a way to get a ring that was both legitimately foreign and cheaply made in Zhejiang.

Wu Zengrong worked on two computers, jumping back and forth between three different accounts. His characters traveled in places with names like Kalimdor, Tanaris, and Dreadmaul Rock; he fought Firegut Ogres and Sandfury Hideskinners. Every once in a while a message flashed across the screen: "You loot 7 silver, 75 copper." Wu couldn't understand any of it; his ex-cook brother-in-law had taught him to play the game strictly by memorizing shapes and icons. At one point, Wu's character encountered piles of dead Sandfury Axe Throwers and Hideskinners, and he said to me, "There's another player around here. I bet he's Chinese, too. You can tell because he's killing everybody just to get the treasure."

After a while we saw the other player, whose character was a dwarf. I typed in a message: "How are you doing?" Wu didn't want me to write in Chinese, for fear that administrators would spot him as a gold farmer.

Initially there was no response; I tried again. At last the dwarf spoke: "???"

I typed, "Where are you from?"

This time he wrote: "Sorry." From teaching English in China, I knew that's how all students respond to any question they can't answer. And that was it; the dwarf resumed his methodical slaughter in silence. "You see?" Wu laughed. "I told you he's Chinese!"

Two months later, when I visited Shifan again, three of the computers had been sold, and Wu was preparing to get rid of the others. He and his friends had decided that playing in Germany was no longer profitable enough; Blizzard kept shutting them down. Wu showed me the most recent e-mail message he had received from the company:

> Greetings,
>   We are writing to inform you that we have, unfortunately, had to cancel your World of Warcraft account. . . . It is with regret that we take this type of action; however, it is in the best interest of the World of Warcraft community as a whole.

The message appeared in four different languages, none of which was spoken by Wu Zengrong. It didn't matter: after spending his twenties bouncing from job to job in factory towns, and having his family relocated for a major dam project, he felt limited trauma at being expelled from the World of Warcraft community. The next time I saw him, he was applying for a passport. He had some relatives in Italy; he had heard that there was money to be made there. When I asked where he planned to go, he said, "Maybe Rome, or maybe the Water City." I stood with him in the passport application line at the county government office, where I noticed that his papers said "Wu Zengxiong." He explained that a clerk had miswritten his given name on an earlier application, so now it was simpler to just use that title. He was becoming somebody else, on his way to a country he'd never seen, preparing to do something completely new. When I asked what kind of work he hoped to find and what the pay might be, he said, "How can I tell? I haven't been there yet." Next to us in line, a man in his early twenties told me he planned to go to Azerbaijan, where he had a relative who might help him do business. I asked the young man if Azerbaijan was an Islamic country, and he said, "I don't know. I haven't been there yet."

After I returned to the United States, I talked with a cousin who played World of Warcraft. He told me he could usually recognize Chinese gold farmers from their virtual appearance, because in the game they stood out as being extremely ill-equipped. If they gained valuable gear or weapons, they sold them immediately; their characters were essentially empty-handed. I liked that image—even online the Chinese traveled light. Around the same time, I did some research on synthetic leather and learned that it's made with a solvent called dimethylformamide, or DMF. In the United States, studies have shown that people who work with DMF are at risk of liver damage. There's some evidence that female workers may have increased problems with stillbirths. In laboratory tests with rabbits, significant exposure to DMF has been proved to cause developmental defects. In other words, virtually everything I had heard from the Lishui migrant workers, in the form of unsubstantiated rumor, turned out to be true.

It was another efficiency of the third-tier factory town. People manufactured tiny parts of things, and their knowledge was also fragmented and sparse. But they knew enough to be mobile and decisive, and their judgment was surprisingly good. An assembly-line worker sensed the risks of DMF; a painter learned to recognize the buildings that mattered; a ring dyer could pick out Sellanyl Yellow. Even the misinformation was often useful—if Christ became more relevant as a Daoist sage, that was how He appeared. The workers knew what they needed to know.

\* \* \*

AFTER I MOVED BACK TO THE UNITED STATES, I BECAME CURIOUS about the small town that Chen Meizi and Hu Jianhui had spent so much time painting. At the Ancient Weir Art Village, I had photographed the artists in front of their work, and now I researched the misspelled signs. All of them seemed to come from Park City, Utah. I lived nearby, in southwestern Colorado, so I made the trip.

I was still in touch with many of the people I had known in Lishui. Occasionally, Chen sent an e-mail, and when I talked with her on the phone she said she was still painting mostly the Water City. The economic downturn hadn't affected her too much; apparently, the market for Chinese-produced paintings of Venice is nearly recession-proof. Others hadn't been so lucky. During the second half of 2008, as demand for Chinese exports dropped, millions of factory workers lost their jobs. Little Long left his plant after the bosses slashed the technicians' salaries and laid off half the assembly-line staff.

But most people I talked to in Lishui seemed to take these events in stride. They didn't have mortgages or stock portfolios, and they had long ago learned to be resourceful. They were accustomed to switching jobs—many laid-off workers simply went back to their home villages, to wait for better times. In any case, they had never had any reason to believe that the international economy was rational and predictable. If people suddenly bought less pleather, that was no more strange than the fact that they had wanted the stuff in the first place. And then in 2009, the Chinese economy regained its strength, and workers made their way back onto the assembly lines.

In Park City, it was easy to find the places that the artists had painted. Most of the shops were situated on Main Street, and I talked with owners, showing them photos. Nobody had any idea where the commission had come from, and people responded in different ways when they saw that their shops were being painted by artists in an obscure Chinese city six thousand miles away. At Overland ("Fine Sheepskim Leather Sine 1773"), the manager became nervous. "You'll have to contact our corporate headquarters," she said. "I can't comment on that." Another shop owner asked me if I thought that Mormon missionaries might be involved. One woman told a story about a suspicious Arab man who had visited local art galleries not long ago, offering to sell cut-rate portraits. Some people worried about competition. "That's just what we need," one artist said sarcastically, when she learned the price of the Chinese paintings. Others felt pity when they saw Chen Meizi who, like many rural Chinese, didn't smile in photographs. One woman, gazing at a somber Chen next to her portrait of the Miners Hospital, said, "You know, it's kind of sad."

Everybody had something to say about that particular picture. The building brought up countless memories; all at once, the painting lost its flatness. The hospital had been constructed to serve the silver miners who first settled Park City, and later it became the town library. In 1979, authorities moved the building across town to make way for a ski resort, and the community pitched in to transfer the books. "We formed a human chain and passed the books down," an older woman remembered. When I showed the painting to a restaurant manager, he smiled happily and said that a critical scene from *Dumb and Dumber* had been filmed inside the Miners Hospital. "You know the part where they go to that benefit dinner for the owls, and they're wearing those crazy suits, and the one guy has a cane and he whacks the other guy on the leg—you know what I'm talking about?"

I admitted that I did.

"They filmed that scene right inside that building!"

When I visited, the Park City mayor kept his office on the first floor of the Miners Hospital. His name was Dana Williams, and he was thrilled to see the photo of Chen Meizi with her portrait. "That's so cool!" he said. "I can't believe somebody in China painted our building! And she did such a great job!"

Like everybody else I talked to in Park City, Mayor Williams couldn't tell me why the building had been commissioned for a portrait overseas. It was a kind of symmetry between the Chinese Barbizon and Park City: the people who painted the scenes, and the people who actually lived within the frames, were equally mystified as to the purpose of this art.

Mayor Williams poured me a cup of green tea, and we chatted. He had an easy smile and a youthful air; he played guitar in a local rock band. "It's the yang to being mayor," he explained. He was interested in China, and he sprinkled his conversation with Chinese terms. "*You mei you pijiu?*" he said. "Do you have any beer?" He remembered that phrase from a trip to Beijing in 2007, when he'd accompanied a local school group on an exchange. A scroll of calligraphy hung beside his desk; the characters read "Unity, Culture, Virtue." He told me that he had first thought about China back in the 1960s, after hearing Angela Davis lecture on communism at UCLA. There was a copy of *The Little Red Book* in his office library. When the Park City newspaper found out, it ran a story implying that the mayor's decisions were influenced by Mao Zedong. Mayor Williams found that hilarious; he told me that he just picked out the useful parts of the book and ignored the bad stuff. "Serve the people," he said, when I asked what he had learned from Mao. "You have an obligation to serve the people. One of the reasons I'm here is from reading *The Little Red Book* as a teenager. And being in government is about being in balance. I guess that has to do with the Dao."

# GO WEST

THE FIRST THING I LEARNED WHILE LIVING ABROAD IS THAT IF you're lost, you have to ask directions. The last thing I learned is that it's possible to ship 143 boxes from Beijing across the Pacific Ocean without a final destination. I've never been good at planning ahead, and this quality became worse after years in China, where everybody seems to live in the moment. And in a country like that it's easy to find a moving agent who's willing to improvise. He went by the English name Wayne, and he wore his hair long, the way Chinese artists often do. When we arranged the contract, Wayne asked my wife, Leslie, if she had any idea where we were going. "It will be a small town, probably in Colorado," she said. "But we haven't decided which one."

"Can you decide within the next few weeks?"

"I think so."

Wayne explained that the shipping container would be on the ocean for much of a month, so the address wouldn't matter, as long as the thing was headed in the right general direction. But after it arrived in the United States, the American partner would need to know where to deliver it by truck. That was Wayne's deadline: we had to find a home in less than five weeks.

Wayne spent two days in our Beijing apartment, managing the moving crew. It consisted of a dozen men, all dressed in clean blue uniforms and carrying metal box cutters. For each piece of furniture, they sliced big squares of cardboard into a size that custom-fit the object. They'd cut off a piece, fold it neatly around the front legs of a chair, and then do the same for the back and the sides. After the cardboard was all taped together, it looked like a chair-shaped box. They created boxes around tables, desks, shelves, stools, and couches. They made something that looked like a giant cardboard bed. An antique three-tiered opium table was perfectly enclosed, layer by layer. It was like watching a team of sculptors work backward, until every object we owned had been converted into a larger, rougher version of itself.

A couple of times, I tried to engage the workers in conversation, but their responses were brief and uninviting. They did not allow us to help. If I picked up an object, somebody immediately grabbed it away, smiling and thanking me profusely. "It's better if they do it themselves," Wayne said, and he was right. They packed the shipping container as tight as a jigsaw puzzle, and a truck carted it off into the night. Suddenly I felt wonderful: all our possessions were gone; we no longer had an address; we could live anywhere we wished. Later that month Leslie and I set off to find a new home.

NEITHER OF US HAD MUCH EXPERIENCE AS ADULTS IN THE United States. I had left after college, to attend graduate school in England, and then I traveled to China; before I knew it I had been gone for a decade and a half. I had never held an American job, or owned an American house, or even rented an American apartment. The last time I bought a car I filled it with leaded gas. My parents still lived in the Missouri town where I grew up, but otherwise nothing tied me to any particular part of the country. Leslie had even fewer American roots: she had been born and brought up in New York, the daughter of Chinese immigrants, and she had made her career as a writer in Shanghai and Beijing.

During the years that I lived in China, I rarely returned to the United States, but I spent a lot of time thinking about the country. Most Chinese were intensely curious about foreign life, and they liked to ask certain questions. What time is it there? How many children are you allowed to have? How much is a plane ticket back? People tended to have extreme views of the United States, both positive and negative, and they became fixated on fantastic details that they had heard. Are American farmers so rich that they use airplanes to plant their crops? Is it true that when elderly parents eat with their adult children, the kids give them a bill for the meal, because they aren't as close as Chinese families? When I taught at a college, a student named Sean wrote in an essay:

> I know that persons in America can possess guns from
> some books and films. I don't know whether it is true. . . . I
> know that beggars must have bulletproof vest from a book.
> Is it true? There is a saying about America. If you want to
> go to heaven, go to America; if you want to go to hell, go
> to America.

It was hard to respond to such combinations of truth and exaggeration. In the early years, it frustrated me, because I couldn't convey a more nuanced perspective. But eventually I realized that the conversations weren't strictly about me, or even about my home country. In China, I came to think of the United States as essentially imaginary: it was always being created in people's minds, and in that sense it was more personal for them than it was for me. The questions reflected Chinese interests, dreams, and fears—even when they discussed America, the conversation was partly about their home.

The longer I stayed overseas, the more I felt something similar happening to my own perspective. China became my frame of reference; I tended to think of the United States mostly in contrast to what I knew in Asia. And my conception of American life became increasingly open-ended. It was hard to envision myself in any particular place, but that also meant that I could live pretty much anywhere. When Leslie and I decided to leave Beijing, both of us had finished researching books, so we knew that our work was portable. We didn't have jobs or children, and we didn't need a long-term home; eventually, we'd probably end up overseas again. And after years of standing out as a foreigner in urban China, I liked the idea of rural solitude and anonymity. A small town in the Rocky Mountains where nobody knew us—that was our own Chinese version of the American dream.

We bought a used Toyota, put a cooler in the back, and followed two-lane highways around Colorado. It was late March and the snow was still deep in the mountains; some of the high passes were closed. At night, we stayed in cheap hotels, and during the day we talked to real-estate agents, who rarely had much to show us. We hadn't realized that middle-class Americans almost never rent their houses; this was before the subprime-mortgage crash and it was easy to buy. In the town of Leadville, an old silver mining community with a population of less than three thousand, I asked an agent if she had anything for rent. "Do you qualify for HUD?" she asked. I said I was pretty sure we didn't; she suggested a mobile home. The only house we saw for rent was a white prefab situated about twenty feet from Highway 24. It was currently occupied by a pack of molybdenum miners, but the real-estate agent assured us the men would be moving out soon; she could put us on a waiting list. Leadville was preparing to open up some mines again, largely because of demand from China. We took a glance at the house and kept driving.

I liked the big bright landscapes, the way the mountains caught the alpenglow in late afternoon, and I liked the heavy-named towns that sat in the valleys: Granite, Bedrock, Sawpit, Crested Butte. In southwestern Colorado we followed the

Uncompahgre River for miles; just seeing that name on a sign made me happy. Not far from the river, a man showed us a brand-new house that sat on an alkali flat. The white soil was as dazzling as broken glass; the thought of writing a book there gave me a headache. Whenever we did find houses for rent, they were usually misfits. They had bad carpet and cheap paneling, or they had been built somewhere in a shaded valley where the snow was slow to melt. Sometimes I sensed that we arrived in the wake of a disaster. Divorces, deaths, bankruptcies—I imagined that was why big houses skidded onto the rental market in small towns.

In a place called Ridgway, we phoned a real-estate agency and happened to talk with a young office manager who had just broken up with her boyfriend. He had left her with the lease on a brand-new house; she hoped to move to Denver and start over. The place was beautiful: high on a mesa, a thousand feet above the Uncompahgre River. From the back, we couldn't see any other houses; the view ran clear across a piñon forest to the fluted walls of the Cimarron Range. Ridgway isn't far from the borders with Utah and New Mexico, and it's home to a little more than seven hundred people. There's one stoplight in the county. Ridgway has no McDonald's, no Walmart, no Starbucks; we couldn't get cell phone reception at the house. It was hard to imagine anyplace more different from Beijing, and we agreed on the spot to a one-year lease.

We bought a futon and some lawn furniture, and camped out on the floor to wait for our shipping container. One afternoon, we drove into the town of Montrose, where we found a couple of wooden bookshelves at an antique market. The dealer agreed to split the delivery fee: we'd pay the first ten dollars and she'd cover the rest. She telephoned her son, who owned a pickup truck. "Twenty-five?" I heard her say. "That's too much. How about twenty?" The Chinese would have appreciated that detail—less than a month back in the States and already I'd witnessed an elderly parent negotiating with her adult child about money.

IN THE EMPTY HOUSE, I SIGNED UP FOR TELEPHONE SERVICE. When I asked for an unlisted number, the phone-company representative said that there would be an extra fee of two dollars per month. For a moment, I weighed my cheapness against the desire for anonymity. "Put it under my wife's name," I said. "Her name is Leslie Chang."

I figured that hers was a relatively common name, but I hadn't thought about how the phone-book listing would appear, with me attached: "Chang, Peter and Leslie." Immediately the mail began to arrive.

> Dear Mr. Peter Chang,
> You love saving money. Better yet, you love saving money and getting better service. So why haven't you switched phone companies?

Leslie and I almost never got anything. Peter Chang was the one, and in the early months he received much of our mail. Credit-card and phone companies sent flyers, as did car dealerships. Peter Chang received advertisements written in Korean Hangul and in traditional Chinese characters. People called at night speaking exotic languages. The Koreans hung up as soon as they realized we didn't understand, but we always tried to keep Chinese telemarketers on the line, to figure out where they were calling from. Who trolled through rural Colorado phone listings in search of Asian names?

For the most part, the callers seemed to be lonely individuals selling long-distance phone cards. But every once in a while a Chinese telemarketer offered something different, and one evening Leslie answered the phone and heard a woman give a pitch for a vacation spot in Wai Er Ming. I listened in, although at first

neither of us could make sense of the name. "Wai Er Ming?"
Leslie said. "Where is that?"

The caller explained that Wai Er Ming is in the American
West, land of cowboys and mountains; the air is fresh and
clean. It was like staring at a puzzle for a few seconds until a
pattern suddenly becomes obvious, and you can't believe you
ever missed it: Wy O Ming.

"Where are you calling from?" Leslie asked. "Are you from
the mainland?"

There was a pause. "Our company is from Hong Kong. But
we do tours in Wai Er Ming."

"I don't believe you're a Hong Kong company," Leslie said.
"A Hong Kong company wouldn't call random people like this.
Also, you don't sound like somebody from Hong Kong. Where
in the mainland are you from?"

On the phone, the woman's voice became very small. "I'm
supposed to say we're from Hong Kong," she said. "I can't tell
you anything else." Afterward I sometimes repeated the word,
just to hear the sound. It had a certain magic, half-strange and
half-familiar: *Wai Er Ming, Wai Er Ming, Wai Er Ming.*

THE SHIPPING CONTAINER ARRIVED LATE. THE DENVER MOVERS
had scheduled delivery for noon on a Tuesday, but their truck
got stuck in snow on Monarch Pass, and then they suffered a
mechanical failure. In our driveway, they backed into a piñon
tree and knocked off a couple of limbs. When the driver realized
he didn't have the key to the container's Chinese-customs lock,
he grabbed a heavy decoupling tool. He grinned and said, "Give
a redneck something to hit with and he'll get it done."

American friends who had moved back from Beijing had
warned us about the feeling you get when your possessions
arrive. It's similar to taking a new baby home from the hospital: all at once you're on your own. In Ridgway, Wayne's
dozen Chinese movers became two Americans named James

and Greg. They did not wear uniforms, and they did not move efficiently. They did not protest when Leslie and I offered to help. The moment they arrived they began to ask about where they could find something to eat. And after James successfully smashed the customs lock, both of them stood in awed silence before the open container.

"I've never seen anything like this," James said finally. "I'll have to tell people about this."

During the rest of the afternoon, while we hauled boxes inside, James and Greg periodically examined the Chinese handiwork. At one point I found both of them crouched in the driveway, studying a table that was completely enclosed in cardboard. "They put us to shame," James said, shaking his head. "This is amazing."

Each box had a number and a label, and James called them out as he went, so Leslie could check off everything on a list. Between boxes, he carried on a running commentary about his Louisiana upbringing, the seven children he and his wife were homeschooling, and the things he had learned in his former life as a long-distance trucker. He had recently gotten rid of his rig because gas prices rose too high. "Sold it to the next guy looking to make a million dollars," he said. "Million problems is more like it." James said that he spent thousands of dollars every year on books, and he told stories about all kinds of topics: trucker fueling strategies, tree nurseries, chicken farms. "They got them on so many drugs nowadays," he said. "I had a friend who worked at a chicken plant, and from the time they're born to the time they're processed it takes eighteen days. Eighteen days! It used to take months. There was one woman who worked there injecting the chickens, and she'd prick herself occasionally by mistake. She got lupus and there was hair growing out of her face. That's why I don't eat chickens anymore. This is box No. 94—office files."

The last thing to be unpacked was our bed, which Leslie had

found years ago in a Shanghai antique market. It had a canopy that consisted of eighteen separate pieces, all carved from elm wood, with intricate scrollwork depicting flowers, human figures, and Buddhist icons. The canopy had no screws, no bolts—only wooden notches and fittings. It had to be assembled in a specific order. We began with one post and worked our way clockwise, with each person supporting a side until the whole thing balanced perfectly. Night had fallen, and the darkness gave the scene a certain intimacy: Leslie and me, James and Greg, all of us together on an early-Republican-era canopy bed surrounded by carved lotus flowers and bodhisattvas and interlocking infinity symbols. After the canopy stood erect in all its glory, James spent a minute studying the fittings. "It's so well designed!" he said. They had six hours of mountain driving back to Denver, but James was cheerful to the end. He shook my hand and wished me good luck; he seemed happy to have another story for the road.

IT WASN'T UNTIL I MOVED BACK TO THE STATES THAT I REALIZED how much I had missed the way Americans talk, especially in small towns. I liked the pacing of their stories, and I liked being able to pick up the nuances of the language. Once when I visited my parents back in Missouri, I took a shuttle bus from the airport, and the driver was a South Carolinian with a huge white beard that tumbled across his chest like a snowdrift. I told him I had been in China until recently.

"Do you speak mandolin?" he asked.

My accent doesn't sound that nice, but I said yes anyway.

"I read a statistic somewhere," he said. "I don't know where, but it said the Chinese could march four abreast into the ocean for all eternity."

The driver talked nonstop for 120 miles. He told stories about his ex-wife, and he described his studies of biblical Hebrew; he had strong opinions about the Book of Daniel. Nowadays he lived in a trailer court in mid-Missouri, but during the 1960s he had traveled to France, Spain, Greece, and Turkey. "I had a rich uncle who took me there."

"Wow, that must have been nice," I said. "What did your uncle do?"

"That was Uncle Sam."

People in China never talked like that. They weren't storytellers—they didn't like to be the center of attention, and they took little pleasure in narrative. They rarely lingered on interesting details. It wasn't an issue of wanting to be quiet; in fact, most Chinese could talk your ear off about things like food and money and weather, and they loved to ask foreigners questions. But they avoided personal topics, and as a writer I learned that it could take months before an interview subject opened up. Probably it was natural in a culture where people live in such close contact, and where everything revolves around the family or some other group.

And a Chinese person with options would never choose to live in a place like southwestern Colorado. The American appetite for loneliness impressed me, and there was something about this solitude that freed conversation. One night at a bar in Ridgway, I met a man and within five minutes he explained that he had just been released from prison. Another drinker told me that his wife had passed away, and he had recently suffered a heart attack, and now he hoped that he would die within the year. I learned that there's no reliable small-talk in America; at any moment a conversation can become personal. When I had DIRECTV installed, a technician came over to drill a hole in the side of the house. He commented that he had just moved to a town called Delta, and I asked him what it was like.

"Quiet," he said. "Not much going on in Delta."

"Why did you move there?"

He looked up from the drill. He was a skinny man in his twenties with blue-line tattoos that ran along his arms like wayward veins. "I had a two-month-old son who died," he said slowly. "That was in Denver, and I just had to get out. I didn't want to stay there any longer. So I moved to Delta."

It took me a moment to respond. "I'm really sorry about that," I said. "It sounds awful."

I didn't know what else to say; in the States, I often had trouble responding to personal stories. But soon I realized that it didn't make much difference what I said. Many Americans were great talkers, but they didn't like to listen. If I told somebody in a small town that I had lived overseas for fifteen years, the initial response was invariably the same: "Were you in the military?" After that, people had few questions. Leslie and I learned that the most effective way to kill our end of a conversation was to say that we were writers who had lived in China for more than a decade. Nobody knew what to make of that; they seemed much more comfortable talking about their most recent prison term.

At times, the lack of curiosity depressed me. I remembered all those questions in China, where even uneducated people wanted to hear something about the outside world, and I wondered why Americans weren't the same. But it was also true that many Chinese had impressed me as virtually uninterested in themselves and their communities. They weren't reflective— they preferred not to think hard about their own lives. That was one of the main contrasts with Americans, who constantly created stories about themselves and the places where they lived. In a small town, people asked very little of an outsider—really, all you had to do was listen.

Sometimes that role made me feel like a foreigner or an impostor, but there was also something comforting about the sense of narrative. It had defined my culture since childhood; even if I was no longer part of the local story, I still understood the way people told it. I liked listening, and I found myself drawn to community events where I could sit quietly in the crowd. Leslie and I went to rodeos and quarter-horse races, where local ranch-

ers competed along with professionals. In the autumn, we attended football games at nearby high schools. We followed tiny
Olathe High through a state championship season, and we went
to the victory parade that was held on Olathe's main street. The
players rode atop fire trucks to the end of the road, where they
did a U-turn and came back, so everybody in town had a chance
to cheer twice.

One weekend in June, we attended a religious rally called
"Cowboy Up for Christ." It was held at the start of rodeo season, and the organizers gave out free copies of *The Way for
Cowboys*, which featured Christian-themed tales from competitive rodeo riders. One speaker was a country musician named
Morris Mott, who talked about growing up in a dysfunctional
family. "When I was sixteen, my personal history met up with
His story, the story of Jesus Christ," he said. He explained how
he had created a different life for himself, and he said that his
faith had helped him cope with the near-death of his child. Mott
had a slow, confident way of talking, and the crowd of two hundred fell silent. "An individual with a story is on a higher ground
than an individual with an argument," Mott said. "Your story is
a powerful weapon you can use, not only against your enemies,
but also to bring other people into the light."

IN THE SPAN OF SIX MONTHS I LOST THIRTY POUNDS. MANY
years earlier, I had been a competitive long-distance runner, but
in Beijing, where the air is badly polluted, I let the hobby go. I
picked it up again in Ridgway, where my home was at an elevation of eight thousand feet, with trails headed off in all directions across the mesa. On runs, I looked for deer and elk and
turkeys; twice I saw mountain lions. I was surprised to find that
I could still run eight or nine miles at a stretch, and soon a light-
ness returned to my legs.

I came to think of this as Peter Chang's healthy period. By
now, his mail was dominated by glossy Chinese flyers for ginseng

products—Prince Gold Heart Formula, Pure American Ginseng Powder—all of them coming from a company called Prince of Peace Enterprises, in Wausau, Wisconsin. Peter Chang also got regular mailings from Korean Airlines. A company called Hellman Motors sent a check for $2,078, along with a letter:

> Attention Peter Chang:
> This Official Notice confirms that you have been selected as a GUARANTEED WINNER in a marketing test for the major automobile companies. This is NOT a joke, prank or gimmick.

I liked it when people pleaded with Peter Chang to accept their money. I imagined him as a lone wolf, a figure of international mystery, and I enjoyed taking his calls. One evening the phone rang just as Leslie and I were returning from dinner in town.

"It's for Peter Chang," Leslie said, after she answered it. "It's a woman. I think she says she's from the National Lightbulb Association."

"What the hell is the National Lightbulb Association?"

"How should I know? Should I just hang up?"

But I decided to hear this one out. The connection was poor, and the woman said something about a one-question poll that would follow a recorded message from Wayne LaPierre, the executive vice president of the association. The message began with an angry voice, and I thought: Man, this French guy seems awfully worked up about lightbulbs! Then it dawned on me that we had confused the word "lightbulb" with "rifle." The NRA was doing a push poll, working the wilds of southwestern Colorado by phone.

LaPierre explained that the United Nations was trying to pass the strictest gun-control treaty in history. Third-world dictators were urging the law forward; it was also supported by liberal American officials and the media elite. After the message, a man got on the phone.

"Mr. Chang," he said, "what's your opinion about these third-world dictators and Hillary Clinton trying to ban firearms in the United States?"

"I'm in favor of it."

"You're in favor of what?"

"I'm in favor of them banning guns," I said. "You have to understand, I'm from one of those third-world dictatorships. I'm from China. I don't think people should have too much freedom."

There was a long pause. "Well," he said. "I appreciate your honesty."

"What did you think I was going to say? If you call anybody named Chang, he's going to say the same thing. We all feel the same way about this. We're all coming from China, and we don't want guns."

"OK," he said. "I understand what you're saying."

"We want a more powerful government, like we have in China."

"Well," he said. "Thanks for answering." He was very polite and he never argued, but he seemed incapable of disengaging himself from this call—not the brightest bulb at the association. At last, I said goodbye and hung up, and Peter Chang took the rest of the night off.

AFTER NEARLY NINE MONTHS IN THE UNITED STATES, LESLIE and I took a road trip to Las Vegas. It seemed like the final act of our homecoming, and we arrived in time for the city's combined marathon and half marathon. Having attended so many rodeos and football games, I decided to make my own return to athletic competition, so I signed up for the half marathon.

The race began before dawn, in front of the Mandalay Bay resort, and the mob of seventeen thousand runners headed straight up the Las Vegas Strip. In a rush, we passed the neon-lit Luxor, the Tropicana, the MGM Grand. Some of the all-night gamblers came outside to cheer. After a couple of miles, I slipped into a faster rhythm; it felt easy, because I had been training at altitude. Soon the race thinned out, and by mile six I led a pack of a few runners, with the next group about fifty yards ahead.

There were professionals in the marathon, Africans and Europeans chasing a $45,000 prize, and they had gone out fast. I knew that somewhere around six miles the half marathoners were supposed to turn off, but I couldn't see anybody up ahead making the break. Finally, I shouted at a bystander in a race-volunteer shirt: "Where are we supposed to turn for the half?"

"Right here," he said.

I skidded to a stop. "Are you sure?"

"Yeah," he said. "You're supposed to go up that street."

The volunteer hadn't been paying attention; he was simply watching the runners go by. But I followed his directions, and not far ahead of me a policeman pulled away from the curb and rolled his lights. And that was when I realized it was the pace car, and I was the leader, and there were more than eight thousand runners following me.

Even when I was young, I had never been good enough to lead a big race. Occasionally, I had won events whose entrants numbered in the hundreds, but anything larger was guaranteed to have athletes who were much better than me. And I knew that today the faster runners were still out there; they had simply missed the turn. If they figured it out quickly, and came back to the course, they'd chase me down without any problem. I promised myself not to look back until mile ten.

In China, I had often dreamed of silence and solitude, but there's nothing quite like the sensation of leading a race. Usually the sport feels visual; you pick out landmarks and athletes ahead, using them as goals. But when you're in front it's all about sound: your breathing becomes distinct, and so does the rhythm of your stride. You listen for footsteps behind you. When a by-stander cheers and then goes silent, you count the seconds until his voice sounds for the next runner.

And I had never imagined how quiet Las Vegas could be. The race continued a few blocks west of the Strip, where the bright lights disappeared and the neighborhood became seedy; I ran by the Las Vegas Community Corrections Center and the Erotic Heritage Museum. I saw a homeless man pushing a shopping cart. He grinned and shouted, "Hey, dude, you're winning!" Rock bands had set up stages along the course, and the musicians were still tuning their instruments. Often they didn't notice me until I was almost past, and they'd try to play something quickly for my benefit. I'd hear the music behind me, growing fainter with the distance, until once again I was alone with my footsteps and my breathing.

At the ten-mile mark I looked back and saw nobody. Soon I was on Frank Sinatra Boulevard, running past the service entrances of the big casinos, and then I reached the finish line in front of the Mandalay Bay. The crowd cheered as I broke the tape; the race director shook my hand. Fifteen minutes later, a Las Vegas television station conducted a live interview with me, along with the winner of the woman's race and the first Elvis to finish—150 competitors had entered the race dressed as Elvis. The fastest one stood proudly with me on TV, dressed in a white Lycra bodysuit with pasted-on sideburns, sweating like the King in concert.

Leslie and I were ushered inside a special VIP tent for the top runners, where we helped ourselves to the breakfast buffet while waiting for the professionals to finish the marathon. One by one, they limped in, mostly Kenyans and Ethiopians with big thighs and whippet-thin calves. They had the haunted look that comes at the end of a long race: gaunt cheeks, empty stares. In the buffet line, a Russian runner looked at me quizzically. "Did you run the race?" she said.

I told her I had won the half marathon.

"You don't look very tired," she said. "You don't look like you ran at all."

She was right—I obviously didn't belong with these athletes. Mine was by far the slowest winning time in the fourteen-year history of the race, and I learned that the lost leaders hadn't realized their mistake until they were already miles off course. (In true Vegas style, a limo took them to the finish line.) The race director assured me that there would be an awards ceremony, but as the morning dragged on, I felt more and more like an impostor sitting in the VIP tent. Finally Leslie and I grabbed a couple of croissants for the road and slipped out.

I never received an award for the race. It was all in the spirit of Peter Chang—he walked away from prizes and free money, and he also knew, like any foreigner, that you have to ask directions if you get lost. In any case, the experience was what mattered most. I had run alone down Frank Sinatra Boulevard, and I had appeared on Las Vegas television. I had shaken the sweaty hand of Elvis himself. Finally I was home, and I had a story to tell; in America that was all you'd ever need.

# DR. DON

In the southwestern corner of Colorado, where the Uncompahgre Plateau descends through spruce forest and scrubland toward the Utah border, there is a region of more than four thousand square miles which has no hospitals, no department stores, and only one pharmacy. The pharmacist is Don Colcord, and he lives in the town of Nucla. More than a century ago, Nucla was named by idealists who hoped that their community would become the "center of Socialistic government for the world." But nowadays it feels like the edge of the earth. Highway 97 dead-ends at the top of Main Street; the population is around seven hundred and falling. The nearest traffic light is an hour and a half away. When old ranching couples drive their pickups into Nucla, the wives leave the passenger's side empty and sit in the middle of the front seat, close enough to touch their husbands. It's as if something about the landscape—those endless hills, that vacant sky—makes a person appreciate the intimacy of a Ford F-150 cab.

Don Colcord has owned the Apothecary Shoppe in Nucla for more than thirty years. In the past, such stores played a key role in American rural health care, and this region had three other pharmacies, but all of them have closed. Some people drive eighty miles just to visit the Apothecary Shoppe. It consists of a few rows of grocery shelves, a gift-card rack, a Pepsi fountain, and a section labeled "Diabetes Supplies," which is decorated with the mounted heads of two mule deer and an antelope. Next to the game heads is the pharmacist's counter. Customers don't line up at a discreet distance, the way city folk do; in Nucla they crowd the counter and talk loudly about health problems. Maybe it's the same instinct that makes people sit close in their pickups, or maybe it's Don, who is always called by his first name, and who seems to have an answer to every question.

"What have you heard about sticking your head in a beehive?" This on a Tuesday afternoon, from a heavyset man suffering from arthritis and an acute desire to find low-cost treatment.

"It's been used, progressive bee sting therapy," Don says. "When you get stung, your body produces cortisone. It reduces swelling, but it goes away. And you don't know when you're going to have that one reaction and go into anaphylactic shock and maybe drop dead. It's highly risky. You don't know where that bee has been. You don't know what proteins it's been getting."

"You're a helpful guy, thank you."

"I would recommend hyaluronic acid. It's kind of expensive, about twenty-five dollars a month. But it works for some people. They make it out of rooster combs."

Next, a women chats about her son, an Air Force officer who has been escorting the bodies of dead soldiers home from Iraq. Another woman inquires about decongestants; a third asks about the risk of birth defects while using a collagen stimulator. Earlier in the day, a preacher from the Abundant Life Church asked about drugs for a paralyzed vocal cord. ("When I do a sermon, it needs to last for thirty minutes.") Another man dropped off a box of reloaded .222 shells for Don. "That's new brass," the man said, setting the bullets on the counter—stiff medicine.

Others stop by just to chat. Don, in addition to being the only pharmacist, is probably the most talkative and friendly person within four thousand square miles. The first time I visited his counter, he asked about my family, and I mentioned new-born twin daughters. He filled a jar with a thick brown ointment that he had recently compounded. "It's tincture of benzoin," he said. "Rodeo cowboys use it while riding a bull or a bronc. They put it on their hands; it makes the hands tacky. It's a respiratory stimulant, mostly used in wound care. You won't find anything better for diaper rash."

DON COLCORD WAS BORN IN NUCLA, AND HE HAS SPENT ALL OF his fifty-nine years in Colorado, where community-minded individuals often develop some qualities that may seem contradictory to an outsider. Don sells cigarettes at his pharmacy, because he believes that people have the right to do unhealthy things. He votes Democratic, a rarity in this region. He listens to Bocelli and drives a Lexus. At Easter, the Colcord family tradition is to dye eggs, line them up in a pasture, and fire away with a 25-06 Remington. A loyal NRA member, Don describes shooting as essentially peaceful. "Your arm moves up and down every time you breathe, so you control your breathing," he says. "It's very similar to meditation." He was once the star marksman of the University of Colorado's rifle team, and for many years he held a range record for standing shooting at the Air Force Academy.

Calmness is one reason why he has such influence in the community. He's short and slight, with owlish glasses, and he seems as comfortable talking to women as to men. "It's like Don looks you in the eyes and the rest of the world disappears," one local tells me. Faith in Don's judgment is all but absolute. People sometimes telephone him at two o'clock in the morning, describe their symptoms, and ask if they should go to the small clinic in Naturita, or call an ambulance for the two-hour trip to the nearest hospital. Occasionally they show up at his house. A few years ago, a Mexican immigrant family had an eight-year-old son who was sick; twice they visited a clinic in another community, where they were told that the boy was dehydrated. But the child didn't improve, and finally all eight family members showed up one evening in Don's driveway. He did a quick evaluation—the boy's belly was distended and felt hot to the touch. He told the parents to take him to the emergency room. They went to the nearest hospital, in Montrose, where the staff diagnosed severe brucellosis and immediately evacuated the boy on a plane to Denver. He spent two weeks in the ICU before making a complete recovery. One of the Denver doctors told Don that the boy would have died if they had waited any longer to get him to a hospital.

At the Apothecary Shoppe, Don never wears a white coat. He often takes people's blood pressure, and he gives injections; if it has to be done in the backside, he escorts the customer into the bathroom for privacy. Elderly folks refer to him as "Dr. Don," although he has no medical degree and discourages people from using this title. He doesn't wear a name tag. "I wear old Levis," he says. "People want to talk to somebody who looks like them, talks like them, is part of the community. I know a lot of pharmacists wear a coat because it makes you look more professional. But it's different here." He would rather be known as a druggist. "A druggist is the guy who repairs your watch and your glasses," he explains. "A pharmacist is the guy who works at Walmart."

He keeps repair tools behind the counter, and he uses them almost as frequently as he complains about Walmart, insurance companies, and Medicare Part D. Since 2006, the program has provided prescription drug coverage for the elderly and disabled, ensuring that millions of people get their medication. But it's also had the unintended effect of driving rural pharmacies out of business. Instead of establishing a national formulary with standard drug prices, the way many countries do, the U.S. government allows private insurance plans to negotiate with drug providers. Big chains and mail-order pharmacies receive much better rates than independent stores, because of volume. Within the first two years of the program, more than five hundred rural pharmacies went out of business. Don gives the example of a local customer who needs Humira for rheumatoid arthritis. The insurance company reimburses $1,721.83 for a month's supply, but Don pays $1,765.23 for the drug. "I lose $43.40 every time I fill it, once a month," he says. The mail-order pharmacies get a better wholesale rate, but Don's customer doesn't like using them; he worries about missing a delivery, and he wants to be able to ask a pharmacist questions face-to-face. "I like the guy," Don says. "So I keep doing it." Since Part D went into effect, Don's margins have grown so small that on three occasions he has had to put his savings into the Apothecary Shoppe in order to keep the doors open.

He is, by the strictest definition, a bad businessman. The pharmacy phone rings; he speaks in friendly tones for five minutes; he hangs up and says, "She sweet-talked me and then she asked me to charge something." The request was for diapers: the answer was yes. He tapes a receipt to the inside wall above his counter whenever something is bought on credit. "This one said he was covered by insurance, but it wasn't," he explains, pointing at a receipt. "This one said he'll be in on Tuesday. This one is a patient who was going on an extended vacation." He counts them off: twenty-three receipts. "The majority of these just don't have the money," he says. Each year he usually writes off between $10,000 and $20,000 in bad credit, and he estimates that in total he is owed around $300,000. His annual salary is $65,000. Over the course of many days in the Apothecary Shoppe, I've never seen a customer walk in whom Don doesn't know by name.

"It's just a cost of doing business in a small town," he says. "I don't know how you can look your neighbor in the eye and say, 'I know you're having a tough time, but I can't help you and your kid can't get well.' It's the number one reason drugstores go under. That was the first lesson in accounting when I was in college."

SETTLERS ORIGINALLY CAME TO THIS REMOTE PLACE BECAUSE they desired an alternative to capitalism. During the 1890s, a group called the Colorado Co-operative Colony hoped to build a utopian community in the region. Its Declaration of Principles complained bitterly about market-oriented competition: "Believing that under the present competitive system only the strong and cunning can 'succeed,' rendering it almost impossible for an honest man or woman to make a comfortable living, and that a co-operative system, if properly carried out, will give the best opportunity to develop all that is good and noble in humanity." (The history of the colony and its values is described in a 2001 dissertation by Pamela J. Clark at the University of Wyoming.)

At the end of the nineteenth century, socialist communities weren't uncommon in the West. Karl Marx had identified the urban working class as most promising for communism, but in America it was often remote regions that became the site of co-operative and socialist ventures. This was partly because of the Desert Land Act of 1877, which offered land at low prices if settlers could find a way to irrigate. Group efforts were necessary to build water systems, and many early communities organized themselves along principles of shared labor. The Mormons were particularly successful—some of their projects in Utah became models for the rest of the West. But most other communities weren't religious; it was far more common to be influenced by the ideas of intellectuals like Marx or Robert Owen. Anaheim, California, was settled through a cooperative water venture, as was nearby Riverside. Others failed but left idealistic names on the map: Equality, Freeland, Altruria.

The Colorado Co-operative Colony chose a desolate region known as Tabeguache Park. It was dry as a bone, but the San Miguel River ran at an elevation only three hundred feet lower, and settlers surveyed the route for an eighteen-mile-long irrigation ditch. Members started digging in 1895. They sold shares all across the country; there were Colorado Co-operative Colony clubs in Brooklyn, Chicago, St. Paul, and Denver. They published a newspaper called *The Altrurian*, which often used the word "comrade" and theorized about cooperation and socialism. They planned to do away with debt, interest, and rent; they banned gambling and the sale of liquor. They dreamed a glorious future: "If a small colony of outlaws and refugees could build Rome and maintain the state for twelve hundred years, who could guess what a well-organized colony of intelligent Americans may accomplish."

Within a year they held their first purge. Ten members were expelled for being too communistic; *The Altrurian* reported that the ditch would be shared, but other property would remain private. During that early period, no matter how bad it got, the newspaper continued to produce a dogged stream of aphorisms. ("Communism may be co-operation, but co-operation is not necessarily communism.") By the winter of 1898, settlers were running out of food. The next year half the board resigned. ("Competition is a product of Hell; Co-operation will make a paradise of earth.") In 1901, the secretary revealed that the colony was bankrupt. A former president committed suicide. ("So long as you think of yourself alone, you cannot be a good cooperator.")

At last, they abandoned the system of shared labor and contracted out to private work crews. In 1904, water finally flowed through the completed ditch; six years later they decided on the name Nucla, after "nucleus." The socialist dreams were never realized, but the irrigation canal continues to function today. And there's still a Colorado Cooperative Company, which employs a full-time "ditch rider" to monitor the system. His name is Dean Naslund, and his father worked on the ditch, and so did his grandfather. Like most Nucla residents,

Naslund doesn't talk about his ancestors in terms of their sociopolitical theories. ("They called him Daddy Joe. He kinda cowboyed. He liked to hop around. Maybe play cards all week sometimes and then work a little.") A local board oversees the water system, but shares are private and disputes can be angry; the founding spirit of cooperation sometimes seems to have evaporated in this dry climate. "Whiskey's for drinking, water's for fighting," Naslund tells me. "I've had a few people get hit with shovels."

Nucla has a reputation as a tough town. It boomed in the 1950s and 1960s, when the region's uranium mining and processing took off. But the uranium market collapsed after the Three Mile Island accident of 1979, and the population continues to drop in Nucla and its sister town of Naturita, which is four miles away. Apart from ranching, the only stable jobs are in the local power plant, a nearby coal mine, and the school system, and there aren't enough positions to go around. In both these towns, the per capita income is less than $14,000 a year, a little higher than half the state figure, and only 8 percent of the population holds a college degree. The school board recently decided to switch to a four-day school week, because of lack of funds. There's only one restaurant in Nucla, one hamburger joint in Naturita, and one bar for both towns. It's called the 141 Saloon, named after the state highway that passes through Naturita. On a Thursday night I'm the only customer, and the bartender tells me that she just bought a three-bedroom house in Nucla for $53,000. That's a mortgage of $250 a month. Her name is Casey.

"Only problem is that the siding is asbestos," she says.

"Is that a big problem?"

"It's not a problem as long as you don't touch it. Asbestos lasts forever." She leans on the wooden bar. "What'll it be?"

"What do you have on tap?"

She smiles and says, "Only thing we got on tap is Jägermeister."

\* \* \*

B<small>Y THE TIME</small> D<small>ON</small> C<small>OLCORD WAS EIGHT YEARS OLD, HE KNEW</small> that he wanted to be a druggist. He grew up in Uravan, a mining town not far from Nucla, and his mother was a clerk in the pharmacy. After school, Don would spend hours in the store, where he liked to watch the druggist at work. As a teenager he began breaking into the place. Along with some friends, he stole beer, *Playboy*s, and condoms. ("The condoms went to waste.") Uravan was a small community, and it didn't take long for the boys to get caught. They weren't allowed to pay back the damages; instead they had to work them off at twenty-five cents an hour. "We had to sweep the store, and stock shelves, and do things like that. Everybody knew why you were there. It was probably the best thing that happened to me."

Around the same time, Don was looking for something in the room that he shared with his brother Jim, and he happened to find a magazine hidden under the bed. It featured photographs of naked men. Many years later, Don would marvel at how a kid living in remote Colorado in the mid-1960s could acquire such a magazine, but this wasn't what went through his mind at the time. He simply thought it was strangest thing he'd ever seen.

"Is this yours?" he said to Jim, when his brother came home.

"Yes," said Jim, who didn't seem embarrassed. He took the magazine back, and neither of them mentioned it again.

Jim was three years older than Don, and he had gotten all the size in the family. He was six foot three and well built, but he had no interest in sports or hunting. Don played baseball and practiced at the rifle range, while Jim liked to go fishing. He spent a great deal of time by himself. In high school he became an excellent student, and he loved being on the debate team. He was a constant source of disappointment to his father, who nagged at Jim to behave like a normal boy. In 1972, a couple of years after Jim left for college, he sent his family a letter explaining that he was gay and that he knew his father would never accept it. He asked them not to look for him; he was leaving Colorado for good. And for the next twelve years nobody heard from Jim.

At the age of eighteen, Don married his high school girlfriend, Kretha, and they lived together in Boulder while he was studying. Eventually, they settled in Nucla and opened the Apothecary Shoppe. In 1983, Don's father died, and one of the first things his widow did was hire a private investigator. The detective found Jim in Chicago, where he was a clerk in the county court. He said he'd had a feeling that something had happened back home.

The following year, Jim made a four-day visit to Nucla. He went for long drives with his mother, who told him that she had always known he was gay, and that she was sorry she hadn't been able to change his father's attitude. In the evenings, Jim and Don sat up late talking. One night, Jim told Don that he had been infected with HIV, and that his doctor said he was likely to develop full-blown AIDS. Jim told Don where he wanted his ashes scattered. And he asked him to visit Chicago, where Jim lived with his longtime boyfriend.

That year, they talked every week on the phone. But whenever the topic of a Chicago visit came up, there was always a reason Don couldn't go. He was too busy at the store; his son and daughter had school activities. Kretha tried to persuade him to make the trip, but he never did.

When Jim died, one of his colleagues telephoned with the news. She sent the ashes in a box, with a copy of his will, some awards from work, and a few photographs. One of the photographs was taken at Wrigley Field, where Jim stands with his boyfriend in front of a "Go Cubs" sign. When Don looked at the photograph, he realized that he knew virtually nothing about his brother. He had seen Jim for all of four days in the past decade; he didn't even know the boyfriend's name. And he understood the real reason he hadn't made a trip to Chicago. "I was angry with myself for not being comfortable in a house where two men were sleeping together," he says. "I didn't want to see two men kissing each other. It wouldn't bother me now, but it did then. I really regret it."

Along with his mother and younger sister, Don scattered Jim's ashes at the juncture of the San Miguel and Dolores rivers. The Dolores flows from the south, where it crosses the great salt dome of Paradox Valley, and the water is saline and has no fish. If you swim there you float as if you were in the ocean, a thousand miles away.

THE LAST DOCTOR IN NATURITA DIED FIFTEEN YEARS AGO. There's a small health clinic, and recently it contracted with a doctor in another part of Colorado to visit two days a week. But the mainstay is Ken Jenks, a physician's assistant who is on call twenty-four hours a day. Jenks has lived in rural Colorado for a decade, and during that time he has learned that electrical tape is harder to remove from a wound than duct tape. Twice he has had patients suffer cervical neck fractures, drive to the clinic, and walk in, when they should have been immobilized at the scene of the accident. "If they had moved the wrong way, they would have been dead or spending the rest of their lives in a wheelchair," he says. It's not unusual for somebody to sign out of the clinic AMA—against medical advice. A couple of times, Jenks has told heart-attack victims that they needed to be evacuated by helicopter, only to have the patients decline because they believed they could get there cheaper. Jenks signed the forms, unhooked the IVs, and the patients got into their pickups to drive the two hours to a hospital. "And they made it," Jenks says. "So they were right!"

Jenks grew up in Salt Lake City, but he has spent most of his working life in small towns. "Maybe I can describe it this way," he says. "I like to play chess. I moved to a small town, and nobody played chess there, but one guy challenged me to checkers. I always thought it was kind of a simple game, but I accepted. And he beat me nine or ten games in a row. That's sort of like living in a small town. It's a simpler game, but it's played to a higher level." Jenks says that he is forced to have "a working relationship" with local methamphetamine users, treating their ailments in confidence. He explains that small towns might have a reputation for being closed-minded, but actually residents often learn to be nonjudgmental, because contact is so intense. "Someday I might be on the side of the road, and the person who pulls me out is going to be a meth user," Jenks says. "The circle is much tighter." He believes there is less gossip than one would assume, simply because so much is already known.

One morning, a young woman arrives at the Apothecary Shoppe after spending the weekend in jail. She had an argument with her husband, who called the police; Colorado state law requires officers to make an arrest whenever they respond to a domestic dispute. The law is intended to protect women from being coerced into dropping charges, but in this case the husband claimed that *he* had been attacked. In the drugstore, the woman is approached by a half-dozen people who have read about the arrest in the local newspaper.

"It's not what it sounds like," she tells one elderly woman. "He's lying about the whole thing, and he's going to get in trouble for that."

"Are you separating?"

"Yes."

They stand at the pharmacy counter. "It's terrible when I have the criminal element in the store," Don jokes. The young woman gets a copy of the newspaper from the front of the store, and she reads the police blotter. "You know what else is crap?" she says. "It says here that it's a second-degree assault, which is a felony. But they dropped it down to a misdemeanor."

"I told my friend you didn't do it."

"He said I attacked him with a frying pan. He said I hit him in the arm. If I'd attacked him with a frying pan, I'd a hit him in the head."

"Let me tell you what you should do," the old woman says. She is in her seventies, with curly white hair and a sweet, grand-motherly smile. "Get you some wasp spray," she says. "It'll put their eyes out."

"I can't even have Mace, because it's a weapon."

With the wisdom of age, the elderly woman explains that wasp spray is not classified as a weapon and is thus available to people who are out on bail. "It's better than pepper spray," she says. "It'll put their eyes out."

A while later, I see the young woman clipping out the arrest listing with a pair of scissors. She doesn't seem embarrassed when I ask why she wants it. "This way if I'm ever stupid enough to think about taking him back, I'll look at this," she says. "I'll keep it in my scrapbook." (Eventually, all charges were dropped, and they divorced.)

At the store, Don never discusses somebody's situation with another person, but he frequently mentions his own problems. Twenty years ago, Kretha was diagnosed with a rare form of spina bifida, and the disease has progressed to the point where she rarely leaves home. For a year the family was unable to get medical insurance, because of her pre-existing condition. Their oldest son flies F-16s for the air force, but their daughter has struggled to find steady work because of alcoholism. After she had difficulties caring for her son, Gavin, Don and Kretha took custody of the boy. Don often mentions such issues to a customer. "If I'm dealing with somebody who has an alcoholic in the family, it helps for them to know about my daughter," he says. "You can't hide anything in a small town. You can't pretend that your family is perfect. My daughter is not perfect, but she's trying." He continues, "Almost all druggists in a small town will tell you the same thing. You are part and parcel of the community. Nobody's better, nobody's worse. We're all the same."

IN NUCLA, WEDNESDAY IS BOWLING LEAGUE NIGHT. THE LOCAL alley shut down to the public long ago, because there are so few people left, but the facility opens twice a week for community leagues. The alley was built in 1962 and all its equipment is original, with an exuberant use of steel that you don't see anymore: long, shiny Brunswick ball racks, dining tables with heavy flared legs. A fifty-year-old Coca-Cola clock hangs on the wall, its hands frozen at a couple of minutes after six. Score cards advertise businesses that have been dead for decades: Miracle Roofing and Insulation, Sir Speedy Instant Printing Center ("Instant Copies While You Wait!"). Don is the league president, and he certifies the lanes every year. He took a course in Montrose in order to be licensed to use a bowling-lane micrometer.

He is always getting certified in order to fulfill some local need. He has taken courses in CPR, and he's licensed to use an electric defibrillator. He has a pyrotechnics display license. He performs regular state inspections on many of the small clinic medicine dispensaries within a hundred-mile radius. He has a pilot's license, and he flies a fifty-year-old Cessna, which he occasionally uses to travel to an inspection. When he heard about a course in California for a certain type of hormone therapy, he flew out and attended two days of classes; and now he compounds medicine for four transgendered patients who live in various parts of the West. He finds this interesting. The patients telephone every three months to order their drugs; he chats and commiserates with them about their hassles with Medicare, which makes it all but impossible for somebody to switch from M to F.

On Friday nights in autumn he announces the Nucla High football games. They play eight-man ball, although if a bigger school comes to town they switch with every possession: when

Nucla is on offense, it's eight-on-eight, but when the other team has the ball it becomes eleven-on-eleven. This is so each side can practice running its offense. Occasionally somebody gets confused, and Don's voice rings over the loudspeakers: "There's eleven white guys and eight blue guys, and that won't work." The football might not be first-rate but the names are a novelist's dream. Nucla has Seth Knob, Chad Stoner, and Seldon Riddle. Dove Creek has a player named Tommy Fury. Blanding has Talon Jack and Sterling Black, Tecohda Tom and Herschel Todachinnie. Shilo Stanley, Terrance Tate, Dillon Daves: if alliteration ever needs an offensive line, recruiting should begin around the Colorado-Utah border.

The Nucla coach is Jim Epright, who is also the principal of the local elementary school. When I visit his office, we chat for a while and he tells me that he grew up in Nucla, moved away for a number of years, and then returned. "Why I came back was that this town took care of me," he says. "In the tough times in my life, there were people here who saw me through it. That's something I can be proud of, as far as being part of this community."

I ask him what he means by tough times. He is a big man with a friendly red face and cool blue eyes. He hesitates for a moment, and then he looks directly at me.

"My mom shot my dad at point-blank range," he says. "He was coming home drunk, and she told him not to come in the house, and he did. He came through the door and she shot him. I was standing ten feet away. And my two brothers were in the next room. Dad was taken out on the Flight for Life, and Mom was taken to jail. You don't have social services in a place like this. I took my brothers to the Ray Motel, and we stayed there a week before I started running out of money."

Here in his office we are surrounded by children's artwork and poems. He continues the story: at the age of fourteen, he moved into the home of "a nice lady who helped other strays";

she supported him while he finished high school. On the weekends he worked in a gas station. He became perhaps the best baseball player Nucla has ever produced, winning a Division I college scholarship. He went off to school and lasted less than half a year: all of his money blown on partying. "I let a lot of people down," he says.

He came home to work in the mines, and then he became a maintenance man for the school system. In his thirties, his supervisor encouraged him to try college again, and he went to New Mexico with his family and got a teaching degree. He has worked in the Nucla and Naturita schools for fifteen years. The elementary school now has 150 students for seven grade levels, and they are losing roughly fifteen kids a year. With each lost student, the state funding drops by about $10,000. The schools recently cut their budget by nearly 8 percent. The starting salary for a teacher is less than $29,000.

"The term 'hanging on by a thread' comes to mind," Epright says. The last time he surveyed the school's students, roughly one-quarter of the children were not living with a biological parent. Many are with grandparents, like Don Colcord's grandson; some live with other relatives or friends. But Epright says that people find a way, and he is impressed by how well the children handle it. "The nontraditional to us seems traditional," he says. "We don't have a foster care system like you'd have in a bigger place."

During summer school, Epright works without pay. He often mows the school lawn; if there's painting to be done, he gets a brush. Last year they needed to cut down some tree branches on the campus, and the bid came in at $10,000, so Epright rented a bucket truck and trimmed the trees with his daughter. He's part of a core group of twenty or so people who seem to hold the community together by volunteering. It's rare to hear somebody mention local government, which has far less impact; Nucla can't even get enough candidates to fill their city council. Epright tells me that teachers, school board members, and local business owners like Don are particularly important. "It's people like that who keep the shallow breathing going," he says.

WHEN OUTSIDERS COME TO TOWN—LONERS, DRIFTERS—THEY often find their way to Don. A number of years ago, a man in his seventies named Tim Brick moved to Naturita and rented a mobile home. He placed special orders at the Apothecary Shoppe: echinacea, goldenseal, chamomile teas. He distrusted doctors, and he often had Don check his blood pressure. It was high, and eventually Don convinced him to get on regular medication. Soon he was visiting every four or five days, mostly to talk.

Don referred to him as Mr. Brick. He had no other local friends, and he was cagey about his past, although certain details emerged over time. His birth name had been Penrose Brick—he was a descendant of the Penrose family, which came from Philadelphia and had made a fortune from mining claims around Cripple Creek. But for some reason Mr. Brick had been estranged from all of his relatives for decades. He had changed his first name, and he had spent most of his working life as an auto mechanic.

One day, his mobile home was broken into, and the thieves made off with some stock certificates. Mr. Brick had never used a broker—to him, they were just as untrustworthy as doctors—so he went to the Apothecary Shoppe for help. Before long, Don was making dozens of trips across Disappointment Valley, driving two hours each way, in order to get documents certified at the bank in Cortez. Eventually, he sorted out Mr. Brick's finances, but then the older man's health began to decline. Don managed his care, helping him move out of various residences; on a couple of occasions, Mr. Brick lived at Don's house for an extended stretch. At the age of ninety-one, Mr. Brick became seriously ill and went to see a doctor in Montrose. The doctor said that prostate cancer had spread to his stomach; with surgery, he might live another six months. Mr. Brick said he had never had surgery and he wasn't going to start now.

Don spent the next night at the old man's bedside. At one point in the evening, Mr. Brick was lucid enough for a conversation. "I think you're dying," Don said.

"I'm not dying," Mr. Brick said. "I'm just going to pray now."

"Well, you better pray pretty hard," Don said. "But I think you're dying." He asked if Mr. Brick needed to see a lawyer. The old man declined; he said his affairs were in order.

Don found a hospice nurse, and within two days Mr. Brick died. Don arranged a funeral Mass, and then he went through boxes of Mr. Brick's effects. There was a collection of old highway maps, an antique cradle telephone, and a Catholic prayer stand. There were many photographs of naked men. Don found checkbooks under four different aliases. There were letters in Mr. Brick's handwriting asking friends if they could introduce him to other men who were "of the same type as me." But he must have lost courage, because those letters were never mailed. Don also found some unopened letters that Mr. Brick's mother had sent more than half a century ago. One contained a ten-dollar bill and a message begging her son to make contact. The bill dated to the 1940s but it still looked brand-new, and something about that crisp note made Don feel sad. Years ago, he had sensed that Mr. Brick was gay, and that this was the reason he had been estranged from his family, but it wasn't a conversation they ever had. Don had figured the old man would bring it up if he wanted to talk about it.

In his will, Mr. Brick left more than half a million dollars in cash and stock to the local druggist. After taxes and other expenses, it came to more than $300,000, which was almost exactly what the community owed Don Colcord. But Don didn't seem to connect these events. He talked about all three subjects—neglecting his dying brother, offering credit to the townspeople, and helping Mr. Brick and receiving his gift—in different conversations that spanned more than a year. He probably wouldn't have mentioned the money that was owed to him, but somebody in Nucla told me and I asked about it. From my perspective, it was tempting to apply a moral calculus, until they added up to a neat story about redemption and reward in a former utopian community. But Don's experiences seemed to have taught him that there is something solitary and unknowable about any human life. He saw connections of a different sort: these people and incidents were more like spokes of a wheel. They didn't touch directly, but each was linked to something bigger, and his role was simply to try to keep the whole thing moving the best he could.

DON COLCORD'S BIRTHDAY IS THE FOURTH OF JULY. THIS IS ALSO when Nucla celebrates its annual Water Days, which commemorates the completion of the town's irrigation system. Each year, there's a parade down Main Street, and Don announces the floats over a loudspeaker and serves as a judge for the best decoration. The winner receives $75. There's a water fight with fire hoses, and a barbecue in the local park, where it's a tradition for a woman to get up onstage and sing "Happy Birthday" to Don. At night he sets off the fireworks, because he's the only person in town who has been licensed to do so.

This year the weather is perfect. In the evening we drive to the top of Nucla hill, where a giant and slightly crooked N has been laid out in white-painted stones facing the town. Behind us, to the west, the sun is setting behind the slate-blue La Sal Mountains. Don is here with his grandson Gavin, and he comments that today's parade was the smallest he's ever seen. This year's theme is Where the Past Meets the Future, and the fireworks, like most of the events, are sponsored by the Lions Club. When Don joined the Lions Club in 1978, he was the youngest member; three decades later, he is still the youngest member.

This weekend he turns fifty-nine. There are only six Lions left, and next year they will decide to close the local chapter.

Some volunteer firemen pull up in a truck, and a couple of ranchers talk idly about hay.

"How's your water look this year?"

"I think it's OK."

"You get your first cutting in?"

"Yep."

They point out the Nucla ditch in the distance—a long straight line, slightly elevated, with the feathered tops of cottonwood trees tracking the route. One of the firemen is Matt Weimar, whose ancestors were among the original settlers of Nucla, and whose family still runs a ranch. He says that not long ago some kid found an old cap-and-ball pistol near the ditch, sitting on a ledge as if some pioneers had just dropped it yesterday.

Trucks and cars arrive from town and park at the bottom of the hill to watch the show. As darkness falls, the Lions prepare the fireworks in metal tubes, and Don ignites them one by one. They raised $1,700 for the display, which is relatively small. But the setting makes it spectacular: reds and blues and greens exploding all around this high hilltop. After it's over, we watch the pairs of headlights glide in a neat line back up Main Street, dispersing as drivers turn off toward home. Our attention drifts upward—now that the fireworks and the headlights are gone, the stars seem brilliant, clustered together like the lights of some faraway city. Don passes around a few bottles of beer. "I don't care if it is a small town, we got good fireworks," Don says. He sips his beer and gazes up at the Milky Way. "When you see them from here, they look so close together," he says. "It's hard to believe they're millions of miles apart."

# ACKNOWLEDGMENTS

THIS BOOK BEGAN WHEN JOHN MCPHEE TOOK AN E-MAIL I HAD written about eating rats in southern China and forwarded it to David Remnick. I'm eternally grateful to John for sending it, and to David for reading it.

At *The New Yorker*, I worked with five editors: Charles Michener, Nick Paumgarten, Dana Goodyear, Amy Davidson, and Willing Davidson. I was very rough when I started, fresh off the Yangtze, and I appreciate everybody's patience. I'm also grateful for the support of Dorothy Wickenden, and for the magazine's wonderful long-distance fact-checking.

If you start at my parents' house in Missouri and search the surrounding two blocks, you will not find a better editor than Doug Hunt. Why keep walking? All of these stories, like all of my books, benefited enormously from Doug's advice. So much talent so close to home.

Fourteen years, one agent, one publisher, one editor, one publicist: does this happen anymore? I thank William Clark, Tim Duggan, and Jane Beirn for being sources of stability in an industry where so much else has changed.

Ian Johnson and Michael Meyer reviewed many drafts, and they've been part of a wonderful community of writers and photographers that I knew in China: Mimi Kuo-Deemer, Travis

Klingberg, Mike Goettig, Matt Forney, Mark Leong, Jen Lin-Liu, Craig Simons, David Murphy. Kersten Zhang, Sophie Sun, and Cui Rong helped greatly with research and fact-checking.

To my parents—thank you for opening your home to so many interesting people during my childhood, and for your examples of curiosity and sympathy. To Leslie—for always, always, always understanding what it takes. To Ariel and Natasha—someday you will read this book, and you will realize how much work was done while juggling multiple diaper changes and night feedings, and you will be inspired to care for me in my old age. For that I am deeply grateful.

A television journalist once said to me, "Are you serious? You ate rat on spec?" All I can say is that a writer has to start somewhere.